Research on Automotive Intelligent Cockpit

The book series presents the latest developments and advances in the multidisciplinary field of intelligent automotive cockpits. The aim is to cover methods (evaluation systems, cognitive models, usability testing, development process, etc.), technologies (system architectures, algorithms, AI and cloud computing, interactive modalities, etc.) and applications (cockpit definition trends, UX/UI design, ecosystems) of the automotive intelligent cockpit as embedded in the fields of engineering, human factors, psychology, design and socioalogy. The series includes monographs, papers and edited volumes in the field of research and development of automotive intelligent cockpit. The book series aims to introduce multidisciplinary knowledge and applications to researchers and engineers. The individual volumes in the series are thematically organised. The aim of each volume is to provide the reader with a comprehensive overview of the design, development or evaluation of the intelligent cockpit in a particular area. As a collection, the series provides valuable resources for a wide audience in academia, engineering research, industry, and anyone else wishing to expand their knowledge of the automotive intelligent cockpit.

Jun Ma · Zaiyan Gong

Automotive Human-Machine Interaction (HMI) Evaluation Method

机械工业出版社
CHINA MACHINE PRESS

Springer

Jun Ma
HVR Lab
Tongji University
Shanghai, China

Zaiyan Gong
HVR Lab
Tongji University
Shanghai, China

ISSN 2948-250X ISSN 2948-2518 (electronic)
Research on Automotive Intelligent Cockpit
ISBN 978-981-99-9950-7 ISBN 978-981-99-9951-4 (eBook)
https://doi.org/10.1007/978-981-99-9951-4

Jointly published with China Machine Press Co.,Ltd.
The print edition is not for sale in China (Mainland). Customers from China (Mainland) please order the print book from: China Machine Press Co.,Ltd.
ISBN of the Co-Publisher's edition: 978-7-111-71594-8

This Springer imprint is published by the registered company Springer Nature Singapore Pte Ltd.
The registered company address is: 152 Beach Road, #21-01/04 Gateway East, Singapore 189721, Singapore

Preface

The automotive industry has undergone unprecedented transformations over the 130 years since the inception of the automobile. Electrification has fundamentally changed how automobiles work, with batteries and electric motors featuring at its technological core, while internal combustion engines and transmissions have begun to fade from prominence. Furthermore, intelligence has transformed the logic behind the assessment of a vehicle's value. Nowadays, software and user experience are the foundation values of new generation vehicles, whereas the number of functions and configurations merely serve as added bonuses. Electrification and intelligence are intertwined trends that are jointly driving the transformation of the automotive industry. As an integral part of automotive intelligence, automotive human-machine interaction (HMI) is at the forefront of this transformation. With the advancement in the Internet of Things (IoT), valuable experience in the development of intelligent vehicles has been gained. The popularization of smart products, such as mobile phones, has refined user habits to intelligent vehicle experiences; and the rapid growth of emerging automotive brands has fostered agile and efficient HMI development and interactive methods. Consequently, traditional brands are drastically adjusting their organizational structure to respond more proactively to this transformation.

Just as ships at sea rely on lighthouses for guidance, so does the development of automotive HMI require a guiding light. However, there is no ready-made solution within the automotive industry. Although various brands are vying for innovation and excellence in the field of HMI, the competition is too close to call, and there is no general consensus on an optimal approach as yet. Smartphones are no longer a sufficient influencing model. Although we have learned from their interaction methods and application ecosystems, intelligent vehicles face unique challenges such as reducing driving distractions, adapting to usage scenarios, and creating immersive spaces, which smartphones do not have to address. Therefore, the only guiding light that can lead the development of automotive HMI lies within the methodology itself. In the face of constantly emerging technologies, scenarios, and demands, automotive companies must discover their own research and development (R&D) methods for HMI. We need product definition methods to understand the requirement of next-generation products, scenario research methods to identify pain points in the user experience, interaction design methods to cater to users' aesthetics and value orientations, software development methods to achieve efficient over-the-air (OTA) updates and iterations as well as HMI evaluation methods to identify product issues and propose suggestions for improvement.

Our focus in this book is on automotive HMI evaluation methods, which aim to reduce driving distractions, lower operational loads, optimize user experience design, and enhance user value.

The content of this book is divided into three parts. Part I, consisting of Chaps. 1–3, introduces the development of automotive HMI and the current status and challenges of its evaluation. We emphasize the industry's need for a comprehensive, systematic, and quantifiable automotive HMI evaluation method. In addition, we propose a three-dimensional orthogonal evaluation system that incorporates all evaluation items into a space matrix composed of three dimensions: interaction tasks, interaction modalities, and evaluation indexes. This enables the proposed evaluation system to achieve completeness and extensibility without overlap when evaluating the complex automotive HMI system.

Part II, comprising Chaps. 4–11, presents a comprehensive elaboration and in-depth discussion on all HMI evaluation indexes. Chapter 4 introduces the origins of seven first-level evaluation indexes and analyzes the differences in user demands between the Chinese and European markets. In Chaps. 5–7, we introduce three rational evaluation indexes: utility, safety, and efficiency, respectively. Each chapter focuses on one first-level evaluation index; it introduces the development of relevant theories, clarifying the associated second-level evaluation indexes, and discussing common issues in HMI design based on actual vehicle testing experience. At the end of Chap. 7, we provide detailed suggestions concerning the selection of suitable interaction modalities for various interaction tasks. In Chaps. 8–11, we introduce four emotional evaluation indexes: cognition, intelligence, value, and aesthetics, respectively. Among them, value and aesthetics are two highly subjective indexes, which previous studies have found challenging to incorporate into a standardized evaluation process. Based on Hofstede's cross-cultural research theory, this book summarizes common differences in value between Chinese and European users. In addition, based on the research on symbolic techniques in interface design, we collate the typical aesthetic orientation of automotive HMI, to standardize the evaluation of these two subjective indexes to some extent.

Part III, consisting of Chap. 12, describes the application of our proposed HMI evaluation system in the automotive R&D process. It elaborates on the methods for using this evaluation system in practice, and the integration of testing and evaluation with the actual product development process to achieve efficient design iterations.

The comprehensiveness and universality of this HMI evaluation system have been verified through a large number of real-time vehicle tests and discussions within automotive companies. The evaluation results can significantly distinguish between automotive HMI systems of different levels, and accurately assess their strengths and weaknesses. It can also fully reflect the differences in demands of users from different countries for automotive HMI systems. By adjusting the weights within the system, it can adapt to users in different markets.

The Human-Vehicle Relationship Lab (HVR Lab) team has long been devoted to research on evaluation methods for automotive HMI. In 2009, the HVR Lab began conducting user interview studies on automotive cockpit functions. In 2014, in collaboration with the Groupe PSA, our team conducted usability tests with a customizable simulated driving cockpit to explore future trends in automotive HMI. We benefited greatly from the HMI research and testing methods of PSA France. In 2018, we developed an HMI evaluation system and testing method for mass-produced vehicles. In collaboration with Banma, we carried out comprehensive evaluations on more than ten well-placed products in the market, and publicly released the test results. The same year, Shanghai AMMI Intelligent Technology Co., Ltd. was established to industrialize intelligent cockpit research and HMI evaluation. After six years of continuous development, this evaluation system has grown in its comprehensiveness and sophistication, evolving into the automotive HMI evaluation method introduced in this book.

While developing this evaluation method, a number of partners from the automotive industry, including Porsche, Groupe PSA, BMW, Volkswagen, Honda, Banma, and Great Wall Motor, among others, have provided us with many suggestions, which have helped us to gradually refine this method. In addition, the intercultural research achievements in non-automotive fields by some experts provided important insights for the development of this method, particularly Prof. Aaron Marcus, Mr. Egbert Schram, and Prof. Paulo Finuras. We would hereby like to express our sincere gratitude to all our partners and collaborators.

Many colleagues from AMMI and HVR Lab were involved in researching this evaluation method and have contributed to the writing of this book. Among them, Liu Dachuan, Lu Jin, Wang Xiaobin, and Hu Fen were intrinsically associated with the development of the evaluation system, while Lu Jin contributed to the writing of Chaps. 8 and 11.

Currently, the automotive intelligent cockpit and HMI industry are thriving, with automotive companies, industry associations, academic organizations, and evaluation agencies all showing keen interest in this area. As one of the earliest research teams in this field, AMMI and the HVR Lab are continuously promoting the application of our evaluation system in industry. Using this evaluation system as a base, the China Automotive Engineering Research Institute (CAERI) has started to implement the testing and certification of the "Intelligent Cockpit Interaction Experience", and the China Association of Automobile Manufacturers (CAAM) has released the "Automotive Intelligent Cockpit Interaction Experience Testing and Evaluation Procedures". With J. D. Power, we co-hosted the second China Intelligent Cabin Award (CICA) in 2023, in which the objective measurement portion was performed using our evaluation system. In the future, our team will further explore this research domain, and make additional contributions to the Springer "Research on Automotive Intelligent Cockpit" book series. We hope these evaluation methods and practical experiences will facilitate the development of automotive HMI in a more orderly, innovative, and sustainable manner.

Shanghai, China
November 2023

Jun Ma
Zaiyan Gong

Contents

About the Authors

Prof. Jun Ma is a professor and doctoral supervisor of School of Automotive Studies, Tongji University, founder of HVR Lab, double-hired professor and doctoral supervisor of College of Design & Innovation, Tongji University, and founder and director of Innovative Design and Entrepreneurship Faculty. He is the deputy director of Intelligent Cabin Division of China SAE, board member of XuanYuan University and judge of XuanYuan Award, and founder of China Intelligent Cabin Award (CICA) with J. D. Power.

He has about 30 years of research and industrial practice experience in automotive product R&D, human-machine interaction, and innovative design. He focuses on the user experience of the automotive industry and improved the academic infrastructure by combining inter-disciplinary cooperation and in-depth industrial practice. He proposed the concepts of "Human-Vehicle Relationship" and "Car Experience Management", established the development process of DDP double-diamond process for intelligent automobile products, and created the C-HVR evaluation system, which is the core content of the cockpit-related group standards of the China SAE.

Professor Jun Ma lived and worked in Germany for 15 years. He graduated from Technical University Darmstadt in Germany with a major in electronic engineering and automation. After graduation, he worked in Germany for Continental, BMW, and Audi with responsibilities in active safety product R&D, project management, Chinese marketing, and strategy.

Dr. Zaiyan Gong is the Director of HVR Lab, Tongji University, General Manager of Shanghai AMMI Intelligent Technology Co., Ltd., Associate Partner of Hofstede Insights, and DUXU board member of HCI International. He holds a Ph.D. of Automotive Engineering from Tongji University.

His current research activities are focused on the definition, analysis, design, and evaluation of the automotive intelligent cockpit. The research covers both rational and emotional fields. In the rational aspect of his work, he created the real-vehicle driving simulation platform, which can do the usability test with objective quantitative data for any mass-produced cars in a virtual driving scenario. In the emotional aspect, he introduced the Hofstede 6-D culture model into automotive UX definition, to show the distinctive user requirement among various target groups.

He has rich experience in automotive industry and worked in Porsche China.

Chapter 1
HMI: An Important Trend in Automotive Development

1.1 Introduction: HMI and Automotive HMI

1.1.1 What is HMI?

Human–machine interaction (HMI) is the study of the design, evaluation, implementation, and other related aspects of interactive machine systems intended for direct use by humans [1].

Human–computer interaction (HCI) is also used in computer science. In theory, the scope of machines is wider than that of computers; accordingly, the scope of HMI is wider than that of HCI. However, pure mechanical interaction systems (e.g., combination locks on safe boxes) are becoming increasingly rare in everyday life and are not the focus of HMI research. Therefore, at present, the scopes of HCI and HMI are essentially the same in both academic research and engineering applications, excluding the need for distinction. In the field of automotive, the term "HMI" is widely used.

HMI is applicable to not only computers, cell phones, and automobiles but also various other fields, including household appliances, industrial equipment, and large interactive facilities in public places. Therefore, the definition of HMI involves a certain vagueness. Nevertheless, for only automobiles, we can provide a precise definition to determine what automotive HMI is and what it is not.

1.1.2 What is Automotive HMI?

Automotive HMI is a system that enables the transmission of dynamic information and emotions between a human and a vehicle, except for the main driving task.

J. Ma and Z. Gong, *Automotive Human-Machine Interaction (HMI) Evaluation Method*, Research on Automotive Intelligent Cockpit,
https://doi.org/10.1007/978-981-99-9951-4_1

Information transmission is the fundamental function of automotive HMI, where the information includes both the instructions input into the vehicle by the driver or passengers and the texts, images, and voice outputs by the vehicle to the driver or passenger. For example, when setting the navigation destination, the information transmitted between the human and the vehicle mainly includes the driver saying "navigate to location X." Then, a list of relevant destinations is presented on the central information display and the driver selects the correct address from this list. Subsequently, navigation commences and the driving direction is shown on the central information display. When designing specific interactions, the information for each step in this example needs to be clarified in a more detailed manner.

Emotion transmission is a relatively new aspect of automotive HMI. Interactions for conveying emotions usually contain little information or no explicit information at all; however, they can express emotions in more complex forms. Such emotions may be the sense of technology, luxury, warmth and comfort, and natural relaxation, as well as more specific and explicit emotions. The main interactions that convey emotions are the in-vehicle dynamic ambient lighting and animations on a screen that create an aura. In other HMI tasks, the interactions for transmitting emotions are usually not emphasized, but their role in automotive HMI is vital and increasingly important.

The information or emotion must be dynamic to meet the definition of automotive HMI. For instance, although a line of words on a door sill plate is considered as information, it is static information; therefore, it does not belong to HMI. The exquisite stitching on the seats, albeit creating a luxurious atmosphere, is also static, and thus, it does not belong to HMI.

Automotive HMI does not include the main and most basic task of driving the vehicle. Specifically, it does not include the position and size of the steering wheel, the steering feel when driving, the foot feel on the clutch and gas and brake pedals, or the feel of shifting gears in a manual-gearbox vehicle. In fact, the main task of driving certainly requires communication between driver and vehicle, and this communication not only determines the driving characteristics of a car but may also affect the driving safety. However, the interaction in the main task of driving is an integral part of the car's dynamics and maneuvering, which is the responsibility of the departments related to the powertrain and chassis in automotive development. There is no need to include this century-old development area in automotive HMI, which is an emerging area in the automotive industry. Furthermore, some functions that are highly associated with the main task of driving are within the vague area of automotive HMI definition. In conventional designs, such functions are usually excluded from the scope of automotive HMI, including gear selection in automatic-transmission models, the turn signal lever on the steering wheel column, and wiper levers. Nevertheless, automotive HMI researchers should also pay attention to these functions if they are to be operated in innovative ways such as knob-type gear shifting and wiper control within the central touchscreen.

According to its definition, the scope of automotive HMI can be extended. A fitting example is an interior ambient light strip with no dynamic effects, which is not included in the scope of HMI; however, once it displays a flowing effect with the navigation direction, it belongs to HMI. In automotive HMI research and design, we should not only focus on conventional buttons, screens, and voice control but also consider how to incorporate more functions and hardware into the scope of HMI.

1.2 Automotive HMI Development Path

1.2.1 Development History

Over the past 100 years, automotive HMI functions have been gradually expanded and their popularity has progressively increased. In the past decade, starting from the 2010s, automotive HMI has experienced unprecedented rapid development and has become one of the most important modules in automotive product design.

Since the 1920s, radio sets have found their way into car cockpits. In 1923, American coachbuilder Springfield and British motor manufacturer Daimler provided original radios for cars at prices of up to approximately 25% of the total vehicle price. Drivers use knobs to adjust the volume and switch channels. In the late 1930s, car radios became interactive, with buttons to select specific pre-defined channels. In 1966, the Ford Thunderbird was the first to include buttons on the spoke of the steering wheel to control the cruising speed, as shown in Fig. 1.1. Subsequently, an increasing number of electronic components have been added to vehicle systems. A case in point: the integration of in-vehicle telephones, radio communication equipment, and satellite navigation devices has realized the electronization and electrification of the cockpit.

Since the 1970s, virtual display units have been widely used in all aspects of public life. In 1986, General Motors launched the Buick Riviera, which was equipped with a touchable central screen, named the "graphic control center", as shown in Fig. 1.2. and the HMI system with a central touchscreen as the control center began to take shape. However, the HMI central control screen was not initially popular due to the inadequate electronic and communication technology at the time. In 2001, BMW unveiled the 7-Series sedan, whose iDrive system was equipped with a non-touch color central screen. It abandoned the design idea of different buttons controlling their respective functions. Instead, all functions were controlled by a single knob, as shown in Fig. 1.3, marking an important milestone for in-vehicle HMI systems.

Automotive HMI's rapid development began in the 2010s. In 2011, the Ford Sync system, which is based on Nuance voice recognition technology, could support over 10,000 voice control commands, representing the transition of voice control from simple mechanical commands to natural language comprehension. In 2012, Tesla premiered its first mass-produced vehicle, Model S, whose central console featured only a 17-inch touchscreen, as shown in Fig. 1.4. This touchscreen integrated more Internet content and replaced the buttons and knobs in the traditional central console, symbolizing the central information display becoming the core of automotive HMI. In 2016, SAIC Motor released the Roewe RX5, which was positioned as "the world's

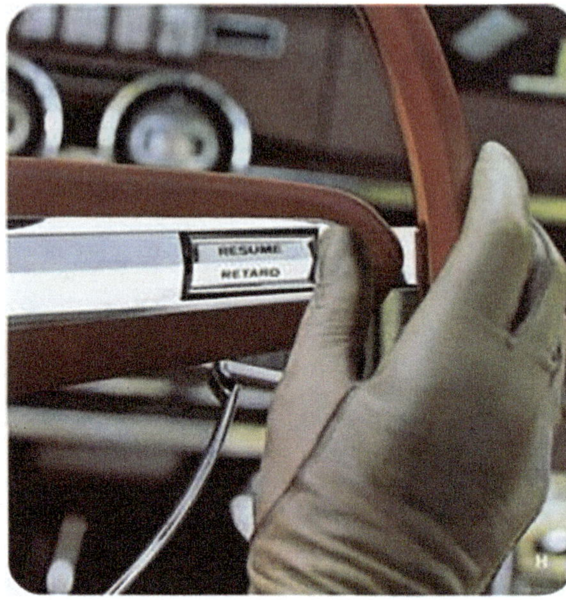

Fig. 1.1 Steering wheel buttons of the Ford Thunderbird (1966) (*Source* Ford Motor Company)

Fig. 1.2 "Graphic Control Center," the central touchscreen in the Buick Riviera (1986) (*Source* General Motors Company)

Fig. 1.3 iDrive system with a combination of knobs and a non-touch central screen in the BMW 7-Series (2001) (*Source* BMW Group)

first mass-produced Internet-connected vehicle." The Roewe RX5 was equipped with the Banma operating system, which was jointly created by SAIC Motor and Alibaba and enabled maps and entertainment content to be online in real time, as shown in Fig. 1.5. This system put forward an interaction framework "maps as the desktop". Its voice control system that could comprehend natural language, and it provided users with OTA upgrade services. Since 2020, HMI has become an important functional module in almost all available automotive products and has been among the most important purchase considerations for consumers. In 2021, among newly released and newly remodeled passenger cars in China, the penetration rate of central touchscreens and voice control was 92.5 and 86.0% [2], respectively. Even low-end models at the 7,000 Euros price range were generally equipped with central information displays and voice control systems.

1.2.2 Mainstream Product Morphology

Automotive HMI development is an extension of human sensory channels. The buttons involve minimal visual and tactile senses, and the central information display increases the information volume transmitted through the visual channel, whereas the voice control utilizes the auditory channel. Such evolution has improved the interaction efficiency, which embodies the people-centered design concept.

Currently, the HMI system of a typical vehicle comprises a central information display, an instrument cluster display, central console buttons, steering wheel buttons, and a voice control system, among other components, as shown in Fig. 1.6. The central information display is the core device in most automotive HMI systems. In

Fig. 1.4 Tesla Model S (2012) with a large 17-inch central information display (*Source* Tesla, Inc.)

Fig. 1.5 Roewe RX5 (2016) equipped with the Banma operating system (*Source* SAIC Roewe)

a hierarchical design, it has the capability of displaying an almost infinite amount of information in its limited area. The easy-to-operate touchscreen serves as both an input and output device. The instrument cluster display, which evolved from the mechanical dashboard of a conventional vehicle, displays maps, navigation, music, and other content in addition to the necessary driving information on traditional gauges. It enables the driver to access such information more efficiently. Central

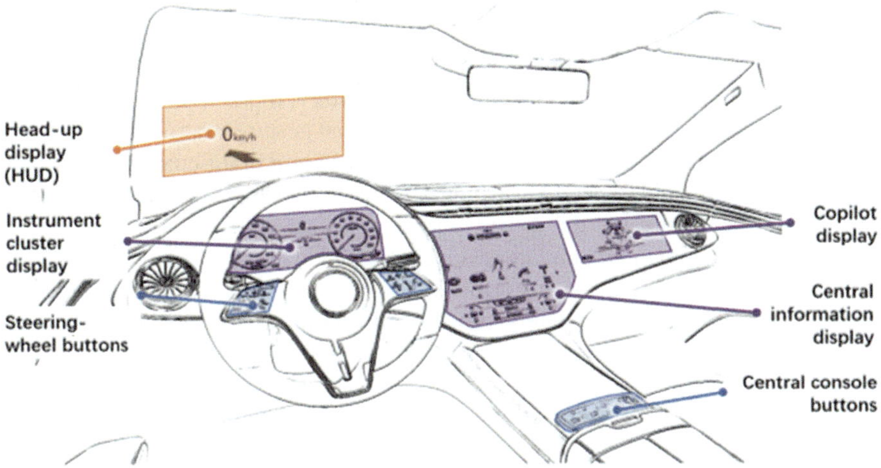

Fig. 1.6 Product morphology of a typical HMI system

console buttons constitute the most important part of traditional automotive HMI; however, as the size and functions of central touchscreens expand, physical buttons have been increasingly replaced by the central display, with some vehicles even completely replacing the buttons in that area. Steering wheel buttons allow drivers to operate by moving only their thumbs without taking their hands off the steering wheel, which enhances driver convenience and safety. Voice control is playing an increasingly important role in automotive HMI. The interaction is more natural because it is very similar to a human-to-human conversation. In some car models, almost all functions in the cockpit can be voice controlled.

Furthermore, the HMI system of some vehicles may include head-up displays (HUDs), lower control displays, and copilot displays. HUDs project information in a virtual image beyond the windshield through light reflection. The driver can read the information without taking their eyes off the road, thus improving driving safety. Via augmented reality (AR) technology, HUDs can superimpose virtual graphics on the road, providing the driver with a richer, more intuitive view of the information directly ahead. Lower control displays are displays below the central information display to replace most or all of the central control buttons. Their hierarchy is simpler, and their operational logic is closer to that of the buttons than to the central display. Copilot displays show information for the copilot, including music, videos, games, among other. As the copilot does not drive the car and can pay full attention to the display, its function and experience is similar to that of a tablet.

In a broad sense, HMI also includes the window actuation buttons on the doors, the seat adjustment buttons, and the light levers on the steering column. These functions are relatively independent and do not differ significantly in design among different car models; therefore, they are not usually the main object of automotive HMI research.

1.2.3 Future Development

Automotive HMI is in constant evolution. In the future, more interaction modalities could be introduced into the vehicle, including spatial gesture control, eye movement control, expression control, and even mind control via a brain-computer interface. Such modalities could expand the boundaries of automotive HMI, presenting new possibilities. Nevertheless, the density and accuracy of information transmitted by these new interaction modalities are relatively limited in the short-term future. Therefore, they will mainly be an expansion of existing HMI systems, instead of completely replacing today's mainstream interaction modalities such as central touchscreens and voice control. In addition, the increasing amount of information related to driving and travel does not mean that the information users need to process should also increase. Scenario-based proactive interaction can simplify this complex issue. Automotive HMI systems should be able to actively filter the options needed by the user based on specific scenarios and present only the filtered options to streamline the interaction experience. This seemingly simpler interface in scenario-based proactive interaction requires powerful information and algorism support.

In essence, automotive HMI development is the constant bridging of the online digital world and the offline real world. Central touchscreens control the physical functions of the vehicle through digital signals inside the car. Through voice interaction, the user's demand for real vehicle control is understood via the digital processing on the cloud. Online maps and ecological services utilize the digital screen to guide users to their destinations in the real world. In the future, artificial intelligence (AI), meta-universe, and other digital concepts will also enter the cockpit to enable users to experience more smoothly and seamlessly across the digital and real worlds.

References

1. Hewett, Baecker, Card, et al. ACM SIGCHI Curricula for Human–Computer Interaction [M]. ACM SIGCHI, 2014.
2. EqualOcean Intelligence. 2021 China Automotive Cockpit Intelligent Development Market Demand Research Report [R]. 2022.

Chapter 2
Overview of Automotive HMI Evaluation

2.1 Development of Automotive HMI Evaluation

Automotive HMI evaluation is an emerging research area mainly based on automotive human factors and usability studies on HMI in non-automotive fields.

2.1.1 Automotive Human Factors

The human factors field is the application of psychological theories in engineering and design. Automotive human factors research focuses on the driver's psychological state, behavior and errors, workload, and trust in technology when driving a vehicle. Its research scope is similar to that of automotive ergonomics, which is more oriented to issues such as posture, field of view, and ease of movement of the human body.

The concept of automotive human factors is relatively new. In the 1990s, its systematic structure came into being in academic research. Based on previous study conducted by McKnight et al., Walker presented the complete Hierarchical Task Analysis of Driving (HTAoD), the most detailed framework for driving task analysis available to date. The highest level of the driving task hierarchy is defined by six first-level sub-goals, which in turn are specified by 1,600 further individual tasks and operations, all bound by 400 tasks and plans containing sub-goal logic operations. Figure 2.1 shows the top-level sub-goal division in HTAoD [1].

Automotive human factors are employed to not only study behavior but also explore the mechanisms for such behavior by using psychological models. Stanton developed a psychological driving behavior model comprising seven factors: feedback, trust, locus of control, mental workload, stress, situational awareness, and mental model. The relationship between these factors is shown in Fig. 2.2. Among them, feedback is received after driver input; trust refers to vehicle performance predictability; locus of control is the degree to which people attribute the cause of an

J. Ma and Z. Gong, *Automotive Human-Machine Interaction (HMI) Evaluation Method*, Research on Automotive Intelligent Cockpit,
https://doi.org/10.1007/978-981-99-9951-4_2

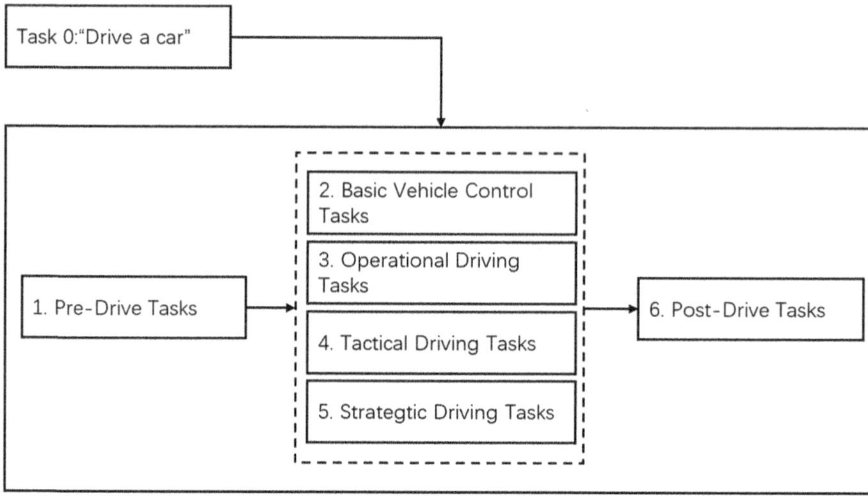

Fig. 2.1 Top-level of the HTAoD (from Walker, 2015)

event to internal or external factors; mental workload represents the occupancy level to the driver's brain power; stress describes the emotional state caused by changes in driving or other aspects of life; situational awareness reflects the driver's activated knowledge for executing the driving task in the transportation system at a given moment; finally, the mental model denotes the model built within the driver for comprehension and reasoning, which sometimes differs from actual situations [2].

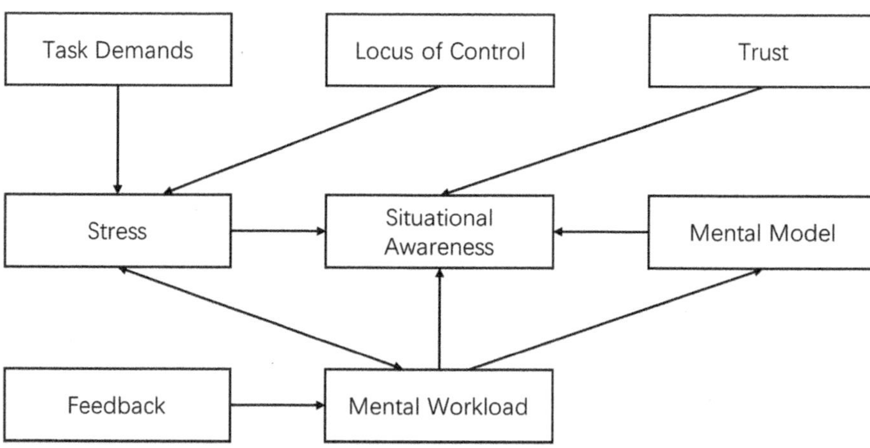

Fig. 2.2 Schematic of the relationship between psychological driving behavior factors (from Stanton, 2000)

Driver distraction is a research sub-field in automotive human factors. As a major cause of road traffic accidents, driver distraction has been extensively studied since the 1960s. Driver distraction sources are complex and varied, and can be voluntary or involuntary, inside or outside the vehicle, and external events or inner thoughts, including eating, talking, using a cell phone (for talking or texting), and using the in-vehicle HMI system [3].

As automotive HMI systems started their burgeoning growth in the 2010s, their research in the field of automotive human factors is in its infancy, and a complete and systematic theoretical framework has not been formed yet. Some early automotive HMI research mainly focused on tasks strongly related to driving itself, such as reading instruments and following navigation guidance, without comprehensively considering the current range of mainstream HMI products.

2.1.2 Usability Evaluation

Similar to smartphones, computers, and home appliances, automotive HMI is also an HMI system, where usability research methods for various HMI systems are equally applicable. In academic circles, the concept of usability dates back to the 1970s. Through continuous refinement of its evaluation methods and applications, usability evaluation has become an important evaluation method for interactive electronic products and systems.

Previous studies have focused on specific usability indexes. Nielsen proposed that usability refers to the ease with which users can use a system's functions, including five specific elements: learnability, efficiency, memorability, errors, and satisfaction [4]. When an interactive system performs well in terms of all five elements, it is considered to exhibit high usability. Among these elements, learnability refers to the users' ability to learn how to use and complete operations faster and with less energy; efficiency means that users can efficiently achieve operational goals; memorability means that the application method of system functions is not easily forgotten; errors refer to the users' error frequency during use; finally, satisfaction refers to the users' subjective feeling and acceptance during use. Hartson suggested that usability contains two aspects: usefulness and ease-of-use. Usefulness refers to whether the functions of a product can be realized, whereas ease-of-use refers to interaction efficiency, learnability, and subjective satisfaction [5]. Product usability as defined by the international standard ISO-9241 refers to the effectiveness, efficiency, and satisfaction that users feel when they use a product in a specific situation to achieve a specific goal [6]. In addition, many other usability index definitions can be found in the literature, as presented in Table 2.1.

In a typical usability test, users answer a subjective questionnaire after using the interactive product and give feedback related to their feelings. The System Usability Scale (SUS) is a commonly used scale [7–9], as shown in Table 2.2. It consists of 10 standard questionnaire items, five of which are positive and five are negative. Each questionnaire item is scored on a five-point Likert scale, ranging from strongly agree

Table 2.1 Usability indexes proposed in the literature

References	Usability indexes
Jakob Nielsen	Learnability, efficiency, memorability, errors, and satisfaction
Rex Hartson	Usefulness, ease-of-use (efficiency, learnability, and satisfaction)
Nigel Bevan/ ISO	Effectiveness, efficiency, and satisfaction
Brian Shackel	Learnability, effectiveness, attitude, and flexibility
Stanton & Baber	Perceived usefulness, task match, task characteristics, and user criteria
Donald Norman	Use knowledge in the world and in the head; simplify task structures; make things visible; get the mappings right; exploit the power of natural and artificial constraints; design for error; standardize
Ben Shneiderman	Time to learn, performance speed, error rate, retention over time, and subjective satisfaction

(5) to strongly disagree (1). Additionally, the NASA Task Load Index (TLX) scale is widely used and includes six subjective questions on mental demand, physical demand, temporal demand, own performance, effort, and frustration. Both the SUS and NASA TLX scale are suitable for the usability testing of products, including automotive HMI systems.

However, HMI during driving involves objective issues, such as driving safety, rather than simply the driver's subjective feelings. Therefore, subjective scales such as SUS and NASA TLX are not sufficiently comprehensive for automotive HMI evaluation, which should incorporate in-depth usability evaluation methods from non-automotive industries and automotive human factors.

2.2 Automotive HMI Testing Methods

2.2.1 Secondary Task Testing

Many common tasks in automotive HMI are secondary tasks, i.e., they require the driver operation on the vehicle while driving. Three main experimental methods are used to study secondary tasks: naturalistic driving studies (NDS), field operational tests (FOT), and driving simulations.

In NDS, the driver operates the vehicle in a real-life driving scenario. The data are recorded during the test without distracting the driver from normal driving to obtain data on the driving behavior that most closely resembles the driver's natural driving. NDS provides a unique opportunity to study driver behavior and performance in the real world, as well as the consequences and risks of actual driving. To avoid disturbing the driver's natural driving and to protect their privacy, sensors and other devices are installed on the vehicle to collect data on vehicle dynamics, vehicle positioning, and

Table 2.2 Standard SUS scale

		Strongly disagree			Strongly agree	
		1	2	3	4	5
1	I think I would like to use this system					
2	I found the system unnecessarily complex					
3	I thought the system was easy to use					
4	I think I would need the support of technical personnel to be able to use this system					
5	I found that the various functions in this system were well integrated					
6	I thought there were too much inconsistency in this system					
7	I would imagine that most people would learn to use this system very quickly					
8	I found the system very cumbersome to use					
9	I felt very confident when using this system					
10	I needed to learn a lot of things before I could get going with this system					

driver behavior, as well as audiovisual recordings of the driving process. NDSs are widely used in driver-behavior-related studies, such as driving behavior classification (aggressive, normal, and cautious driving) [10], driving distraction studies [11], fatigue driving, drunk driving studies [12], and vehicle-following model evaluation [13]. The main advantage of NDS is their ability to obtain long-term natural driving behavior data of drivers, e.g., over months or years, to eliminate the possibility of the driver adjusting their behavior due to feeling observed or being in a test situation.

In studies that include automotive HMI as a secondary task, there are limitations in the NDS approach. First, the driver is not subject to numerous constraints during naturalistic driving and may not be sufficiently focused on completing the driving task. Furthermore, other traffic participants might cause interference with the driver's behavior in an actual complex driving environment. Both of these uncertainties cannot be completely eliminated; therefore, identifying the driving interference caused by HMI independently is difficult. Therefore, in most NDS-based secondary task studies, HMI is usually considered as one of the secondary tasks without being divided and analyzed in depth [14]. Second, NDS cannot include complex tasks that may negatively affect driving safety as they are restricted by driving safety and ethics, which limits their research scope [15]. Third, NDS is usually not equipped with sophisticated data collection equipment to avoid interfering with the drivers' natural driving, which results in a limited data diversity, making it impossible, for example, to monitor physiological data (e.g., drivers' eye movements) or locate vehicles at the

Fig. 2.3 Vehicle equipped with high-precision positioning equipment at a closed testing site

decimeter or centimeter level. Moreover, NDS testing involve massive amounts of data over long periods of time, requiring considerable time, manpower, and capital investments.

FOT is a large-scale field tests carried out to evaluate the efficiency, quality, robustness, and acceptance of automotive-related solutions, such as navigation, traffic information, and driver assistance systems. FOT generally employs a dedicated driving route rather than free driving and are conducted at a professional closed test site if conditions allow. The test scenarios and cases should follow detailed test protocols to achieve testing process standardization. Vehicles for FOTs can be equipped with sophisticated data acquisition equipment, such as eye trackers, detection-response task (DRT) testing equipment [16], and high-precision positioning equipment, to acquire richer data than those involved in NDS, as shown in Fig. 2.3.

In studies that include automotive HMI as a secondary task, FOT, with its standardized testing procedure and various test data, is able to analyze driver distractions caused by automotive HMI and trace the root cause of each distraction, so that problems in HMI design can be accurately determined. However, FOT is highly demanding in terms of the testing field and procedure management. Without an appropriate test design, factors such as sunlight reflection angles and road roughness may influence the test results. Therefore, the time and monetary costs of FOT under ideal conditions are high. In addition, FOT generalize and standardize the test cases, instead of fully restoring the real driving scenario, so they cannot completely substitute NDS in novel or lengthy continuous scenarios.

In driving simulation, drivers operate virtual vehicles in a relatively static environment via computer simulations and image displays. They are frequently used to study driving behavior and secondary tasks [17, 18]. The driver's control of the virtual vehicle is achieved through a specialized steering wheel and pedals. Hardware devices transmit data to the computer responsible for scenario operation, which then feeds the corresponding scenario changes back to the driver via video, audio, vibrations, and other stimuli based on the input data. Various kinetic data during vehicle maneuvering can be read directly from the driving simulator software, and devices such as eye trackers and cameras are often used to record the driver's physiological state. Driving simulators can be classified into three categories according to equipment complexity and the immersiveness of the experience. A simple driving simulator consists of a monitor, computer, steering wheel, pedals, and interactive screens, usually without a complete cabin structure and interior design. A static-cabin driving simulator is modified from a real vehicle cabin and uses the cabin structure of the real vehicle, including the cockpit, interior panels, steering wheel, seats, and usually its body and doors as well. The virtual driving environment is typically projected on a large-sized curve screen to achieve panoramic immersion, as shown in Fig. 2.4. A multiple degree-of-freedom (DOF) panoramic driving simulator comprises a multi-DOF simulator cabin that houses a cabin modified from an real vehicle and a near-360° panoramic ring screen, as shown in Fig. 2.5. The simulator cabin tilts and moves as the vehicle accelerates, decelerates, and turns, allowing the driver to experience an acceleration feeling akin to that of real road driving. Recently, the development of virtual reality (VR) helmet technology has ushered new driving simulation methods, however, VR-helmet-simulated driving has not yet been widely applied in scientific research.

Driving simulations have many advantages. First, drivers can be absolutely assured of test safety by driving vehicles in simulated scenarios where no actual accidents occur. Tests that have potential safety hazards, which often cannot be performed in FOT or NDS, can be conducted on driving simulators. Second, driving simulations enable flexible designs of parameters, including road width, lane number, road gradient, road pavement type, road adhesion coefficient, traffic density, and weather conditions, with high flexibility according to test requirements. Standard road scenarios avoid biased test results caused by road wear, weather changes, and

Fig. 2.4 Static-cabin driving simulator

Fig. 2.5 Panoramic 8-DOF driving simulator at Tongji University

event trigger timing, among other factors, in real-world road environments. Furthermore, driving simulations are highly efficient for testing, time saving for data acquisition equipment installation, and commissioning in FOT, as well as the time spent going back and forth between different routes. Regardless, there are two common issues in driving simulation tests. One is the unreasonable monitor position and screen settings, resulting in significant differences between the perspective occupied by similar objects in the screen and that in a real driving scenario, which affects the driver's perception of speed during a driving simulation [19]. A perspective that is too small diminishes the driver's sense of speed, which in turn reduces the difficulty of the driving task. To restore a realistic view, driving simulators usually require the use of large-angle projection curve screens or a side-by-side arrangement of three to seven large displays. The other issue is the dizziness experienced by subjects in driving simulations. During driving, the driver's body is at a static state, while the scenes perceived by the eyes are in motion and vary, and such contrast can cause dizziness. A multi-DOF simulation cockpit can relieve some of this dizziness by creating a sense of acceleration. The level of dizziness in a driving simulation varies greatly among individuals. Most people can no longer feel noticeable dizziness after a 10–20 min adaptation process.

In studies that include automotive HMI as a secondary task, driving simulations are suitable for most basic research. However, when evaluating real vehicles, conventional driving simulations encounter certain problems [20]. For instance, the cockpit in a driving simulator is usually fixed and non-interchangeable, so its interior layout is different from that of the target test vehicle. To restore the test vehicle's HMI system, the design angle, relative position, and other ergonomic parameters of the cockpit screen and related devices must be adjusted. However, limited by the cockpit interior structure and modification costs, these restorations often do not yield ideal results.

2.2.2 Real Vehicle Driving Simulation Bench

To comprehensively and efficiently test the HMI system of real vehicles, the research team at the Human-Vehicle Relationship Lab developed a real vehicle driving simulation bench, as shown in Fig. 2.6. The developed simulation bench can not only collect rich real-time driving behavior data but also fully considers the HMI layout and design factors of real vehicles in the evaluation. The real vehicle driving simulation bench consists of the following five major components:

- Curve screen: With a diameter of 7 m and an angle of view of approximately 240°, its images are projected by three high-definition projectors. The large-sized curve screen allows drivers to better immerse themselves in the simulated driving environment, and the driving performance is closer to that on real-world roads.
- Simulated driving environment: The simulated driving environment includes virtual roads, vehicle dynamics models, and traffic flow. The road design is based on real road scenarios and is designed in compliance with specific test items and procedures. To ensure realistic vehicle dynamics and better cooperation among the various hardware and software in the entire system, SCANeR Studio was used; it is a proven and widely used driving simulation software.
- Quick connection system: Any mass-produced passenger car can be connected to the simulated driving environment quickly (within dozens of minutes), allowing the driver to sit in the cabin of an real vehicle and drive it in the virtual environment using the original steering wheel and pedals of the real vehicle. A specialized device under the front wheels of the vehicle captures the steering angle of the wheels and provides force feedback. Sensors beside the accelerator and brake

Fig. 2.6 Real vehicle driving simulation bench in the Human-Vehicle Relationship Laboratory

pedals capture the pedal travel and pressure, which is fed to the simulator software, so as to control the vehicle in the virtual scenario.
- Data acquisition system: Vehicle, road, and physiological data can be captured in the developed simulation bench. Vehicle data, including acceleration, braking, and steering angle, are obtained from the quick access system. Road data, including driving speed, lane departure, and distance ahead, are obtained from the driving simulation software. Physiological data are the dynamic data of the eyeball obtained from a head-mounted eye tracker, including fixation position and fixation duration, among others.
- Evaluation management system: The evaluation management system can synchronize, integrate, and analyze all types of data collected, and score each vehicle test item based on the embedded evaluation system.

Testing automotive HMI with a real vehicle driving simulation bench has many advantages. First, the subject vehicle is a real mass-produced vehicle, instead of a modification based on a fixed cockpit; thus, the interactive device layout of the subject vehicle need not be restored, as is the case in a conventional driving simulator. Second, the data collection range is comprehensive and the process is simple. Particularly, the test vehicle coordinates are obtained directly from the simulation driving software; thus, no special data collection is required, not to mention positioning deviation. Third, standardized tasks and events can be designed in the simulated driving environment and precisely triggered by the computer, which avoids the random occurrence of events in NDS. Fourth, environmental variables can be controlled well in the simulated driving environment to avoid the potential impact of sunlight angle and rough roads on driver behavior in FOT. Finally, as opposed to FOT, driving simulations do not require frequent transitions and scheduling, which can improve the test efficiency by approximately 3–5 times.

During use of the real vehicle driving simulation bench, attention must be paid to the restoration of vehicle dynamics and driver's perspective. Vehicles connected through the quick connection system should have a similar steering feel, pedal feel, acceleration performance, and braking performance in the simulated driving environment as those on real-world roads; otherwise, the test results may differ from the real vehicle performance. The vehicle is supposed to be parked where the driver's seat is near the center of the ring screen, thus allowing the driver to observe a distortion-free picture of the road scenario. Meanwhile, the horizon height and lane width should also be carefully adjusted to achieve the most authentic driving experience.

2.2.3 Eye Tracker

Eye trackers are essential data acquisition devices in both FOT and driving simulation. Human eyes are an important medium to acquire information from our surroundings and transmit images in our sight to our mind in real time. The eye-mind hypothesis states that when there is a visual target in the mind, the position of eye gaze is usually related to the target being focused on and thought about [21]. Eye studies can obtain information in multiple layers, with the surface layer being "what is seen" and the deep layer "what is thought." For richer visual information, humans have evolved a series of eye-movement behaviors, including saccade, fixation, and pursuit movements. Saccades are rapid, ballistic eye movements that abruptly change the point of fixation. Fixation is the process of holding the visual center at a point for a certain period of time to acquire image information. As the eyes are in a state of rapid movement during the process of saccade and its duration is extremely short, the vast majority of visual information is acquired during the process of fixation. Pursuit movements refer to the eye-movement behavior in which the observer and the object are in relative motion, and the observer keeps observing the object while maintaining it in the central visual field. To capture and record the above eye-movement behaviors, a professional eye tracker is required.

Eye trackers also play an important role in basic psychological research. Eye trackers are used to record the track features of eye movements when processing visual information, and are widely applied in attention, vision, cognitive psychology, and other related research fields. Eye trackers can extract the fixation point, fixation duration and counts, saccade distance, and pupil size, among other features, from the trajectory record, so as to study people's internal cognitive process. Eye movement studies have evolved through various methods, including the observation, afterimage, mechanical recording, optical recording, and image recording methods. Since the 1960s, rapid advances in camera technology, infrared technology, and computer technology have promoted the research and development of high-precision eye trackers. The structure of modern eye trackers generally includes four systems, namely, the optical system, pupil central coordinate-extraction system, vision and pupil coordinate-superposition system, image and data recording and analysis system.

According to their working principle, eye trackers can be categorized into current recording eye trackers, electromagnetic induction eye trackers, image/video eye trackers, and pupil-corneal reflex eye trackers [22]. Currently, mainstream eye trackers on the market generally adopt the pupil-corneal reflex working principle.

According to the application scenario, eye trackers can be categorized into wearable and screen-based eye trackers, as shown in Fig. 2.7. Wearable eye trackers integrate an eye-movement acquisition device and camera on lightweight glasses to collect the user's eye-movement behavior in a real environment. During use, users can move freely and interact with the environment naturally, so the experimental results are close to those in the real scene. Wearable eye trackers are capable of recording scenes seen by the users and recording eye-movement behavior during observation. Screen-based eye trackers incorporate an eye-movement acquisition device and a display device, which are placed together at a certain distance from the user to record the user's eye-movement behavior while observing the screen. Wearable eye trackers are commonly used in studies of automotive driving behavior and automotive HMI as they are capable of capturing a wider field of view (FOV for short) and are more suitable for different vehicle cockpits. Eye tracker application in automotive HMI studies is further discussed in Sect. 6.2 of Chap. 6.

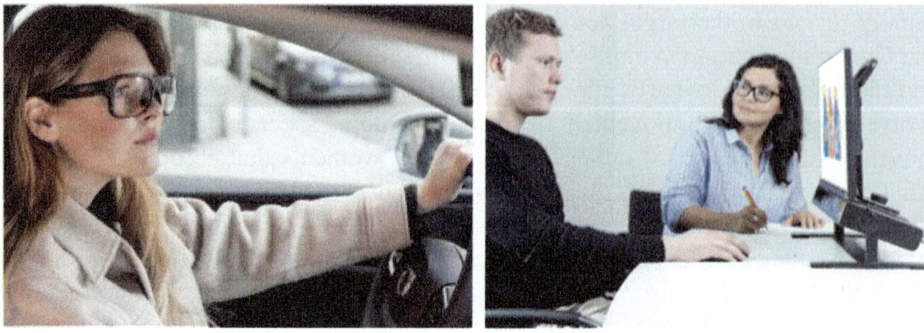

Fig. 2.7 Tobii pro glasses wearable eye tracker and Tobii pro spectrum screen-based eye tracker (*Source* Tobii)

References

1. Walker G, Stanton N, Salmon P. Human Factors in Automotive Engineering and Technology [M]. CRC Press. 2015.
2. Stanton A, Young S. A proposed psychological model of driving automation [J]. Theoretical Issues in Ergonomics Science, 2000, 1, 315–31.
3. Parnell K, Stanton N, Plant K. Driver Distraction [M]. CRC Press. 2019.
4. Nielsen J. Usability Engineering [M]. San Diego, California: Academic Press, 1992.
5. Hartson H R. Human–computer interaction: Interdisciplinary roots and trends [J]. Journal of Systems and Software, 1998, 43 (2): 103–118.
6. Ergonomics of human-system interaction - Part 210: Human-centred design for interactive systems: ISO 9241-210:2019 [S].
7. ZF Bangor, A., Kortum, P. T., & Miller, J. T. An empirical evaluation of the System Usability Scale [J]. International Journal of Human–Computer Interaction, 2008, 24, 574–594.
8. ZG Sauro, J., & Lewis, J. R. Quantifying the user experience: Practical statistics for user research (2nd ed.) [M]. Cambridge, MA: Morgan Kaufmann, 2016.
9. Schmitt, N. Uses and abuses of coefficient alpha [J]. Psychological Assessment, 1996, 8 (4): 350–353.
10. Feng F, Ba O S, Sayer J R, et al. Can vehicle longitudinal jerk be used to identify aggressive drivers? An examination using naturalistic driving data [J]. Accident Analysis and Prevention, 2017,104: 125–136.
11. Ye M, Osman O A, Ishak S, et al. Detection of driver engagement in secondary tasks from observed naturalistic driving behavior [J]. Accident Analysis and Prevention, 2017,106: 385–391.
12. Al-Sultan S, Al-Bayatti A H, Zedan H. Context-aware driver behavior detection system in intelligent transportation systems [J]. IEEE Transactions on Vehicular Technology, 2013, 62(9): 4264–4275.
13. Tawfeek M H, Basyouny K E. A perceptual forward collision warning model using naturalistic driving data [J]. Canadian Journal of Civil Engineering, 2018, 45(10):899–907.
14. Young K L, Osborne R, Grzebieta R, et al. Using naturalistic driving data to examine how drivers share attention when engaging in secondary tasks [J]. Safety Science, 2020, 129: 104841.
15. Santos J, Merat N, Mouta S, et al. The interaction between driving and in-vehicle information systems [J]. Transportation Research Part F Psychology & Behaviour, 2015, 8(2): 135–146.
16. Road vehicles — Transport information and control systems — Detection-response task (DRT) for assessing attentional effects of cognitive load in driving: ISO 17488:2016 [S].
17. Xiaomeng L, Atiyeh V, Sebastien D, et al. Effects of an in-vehicle eco-safe driving system on drivers' glance behaviour [J]. Accident Analysis & Prevention, 2019,122: 143–152.

18. Grahn H, Kujala T. Impacts of Touch Screen Size, User Interface Design, and Subtask Boundaries on In-Car Task's Visual Demand and Driver Distraction [J]. International Journal of Human-Computer Studies, 2020, 142:102467.
19. Palmer, Stephen E. Vision Science: Photons to Phenomenology [J]. Quarterly Review of Biology, 1999, 77(4):233–234.
20. Metz B, Schoch S, Just M, et al. How do drivers interact with navigation systems in real life conditions? [J]. Transportation Research Part F Traffic Psychology & Behaviour, 2014, 24(5):146–157.
21. Aga Bojko. EYE TRACKING THE USER EXPERIENCE, A Practical Guide to Research [M]. Beijing: People's Posts and Telecommunications Press, 2019.
22. Yan Guoli, Bai Xuejun. Foundation and Application of Eye Movement Analysis Technology [M]. Beijing: Beijing Normal University Press, 2018.

Chapter 3
Structure of Automotive HMI Evaluation Systems

3.1 Challenges in Automotive HMI Evaluation

Automotive performance evaluation is of great importance for engineering research and development as well as consumer purchasing choices. Despite automobiles being highly complex systems, almost every performance can be simplified into a few evaluation indexes, which are more convenient to operate and easier to communicate. For example, car power performance can be evaluated via engine horsepower, 0–100 km/h acceleration time, and maximum speed. The fuel economy can be evaluated in terms of its fuel consumption per 100 km. The handling stability can be evaluated via slalom and moose tests, among others.

The evaluation of these performances can be simplified because each performance has an explicit objective and the operation method is determined. For example, the core goal of power performance is acceleration. Although maximum speed, gradeability, and tractive capacity are also power goals, the performance is positively correlated with acceleration; thus, special measurements are not required usually. The only way to accelerate is to push the accelerator deep and delay the shift as much as possible.

However, in automotive HMI, none of these seemingly natural rules apply. The task goals of automotive HMI are not single, but complex, discrete, and weakly correlated to each other. Automotive HMI evaluation faces the following challenges.

Numerous Interaction Tasks

The information displayed on the central information display and the functions operated on the instrument cluster display of a vehicle usually exceed 1000 items. Evaluation based on partial information and tasks may not reflect the actual condition of the entire interactive system. For example, an interactive system suitable for searching music may not be well suited for searching navigation destinations. Even for the very specific task goal of searching for a navigation destination, a system that is convenient to type in text may fail to find the place being searched for.

J. Ma and Z. Gong, *Automotive Human-Machine Interaction (HMI) Evaluation Method*, Research on Automotive Intelligent Cockpit,
https://doi.org/10.1007/978-981-99-9951-4_3

Covering such a large number of functions in a set of evaluation systems is challenging; therefore, important tasks should be filtered for key evaluation, and the evaluation indexes and testing methods may vary among important tasks of different types.

System, Software, and Ecological Integration

Although automotive HMI systems are similar to smartphones in terms of function and logic, they have a larger evaluation scope than smartphones.

When evaluating a smartphone, the focus is on its operating system (e.g., page layout, shortcut menus, and on-screen gestures) rather than on specific applications (e.g., Apple Maps, Spotify, and WeChat). This is because these common applications have almost identical interaction and performance on all major smartphones and do not require special evaluation. However, for the HMI system of each vehicle, in addition to the different operating systems, there are also differences in the interface design and interaction logic for their navigation, music, and communication functions. Therefore, automotive HMI system evaluation must consider these specific functions or the application software.

Furthermore, as the automotive HMI system accesses an increasing number of online ecological resources, the richness of these resources also needs to be evaluated. For example, despite the excellent interaction design of a vehicle's music software, users will not use it very often if the online music library fails to offer currently popular songs.

Various Interaction Modalities

There is a variety of operating modes for automotive HMI. The same interaction task can often be achieved by multiple interaction modalities, which makes the evaluation object no longer a linear catalog, but an interlaced matrix, thus requiring a more complicated scoring logic. For example, a navigation destination can often be input using various interaction modalities, such as keystroke typing, touchscreen handwriting, and voice control. A vehicle with excellent touchscreen handwriting recognition may fail to recognize speech. Therefore, whether the vehicle's navigation destination input performance is good or poor requires specially designed computational methods.

Large Proportion of Secondary Tasks

Unlike most interaction systems, many tasks in automotive HMI require user operation while driving instead of concentration on the interaction. These interaction tasks are called secondary tasks. Secondary task execution should not carry considerable driver distraction, otherwise it will have a negative impact on driving safety.

When evaluating secondary tasks, the user is required to drive the car on an real road or driving simulator, which has an impact on the secondary task itself. For example, a user may take 10 s to enter a navigation destination when the vehicle is stationary; however, they may take double that time while driving owing to the driving safety consideration.

Moreover, the impact of secondary tasks on driving needs to be evaluated. For example, a significant lane departure while driving due to entering a navigation destination could result in a collision with a vehicle in the adjacent lane; if the driver loses sight of the road for two consecutive seconds, the user may not be able to react in time when the vehicle ahead brakes in an emergency.

3.2 Structure of Three-Dimensional Orthogonal Evaluation Systems

The comprehensive evaluation of automotive HMI is a complex topic. The evaluation system must be well structured so that hundreds or thousands of measurements can be arranged in an orderly manner without overlap or gaps. Additionally, the function and operation modes of automotive HMI will develop rapidly with time, and this evaluation system needs to be able to expand while maintaining the stability of the original structure.

The automotive HMI evaluation system should be a three-dimensional matrix space, including interaction tasks, interaction modalities, and evaluation indexes (Fig. 3.1), which are orthogonal, independent, and do not affect each other.

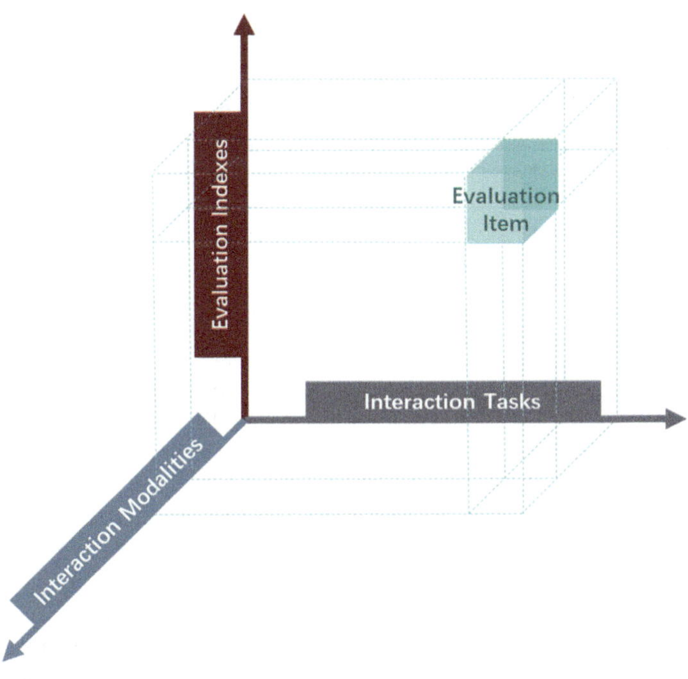

Fig. 3.1 Structure of the three-dimensional orthogonal evaluation system

3.2.1 Interaction Tasks

Interaction tasks refer to a set of activities that the driver performs to achieve a certain goal when using the automotive HMI system. In a broad sense, these activities can be either physical movements, or perceptual or cognitive activities. However, in general, interaction tasks are sometimes in a narrow-sense scope, that is, they require the driver to perform physical activities or verbal dialogues to carry out an operation, along with its corresponding output and feedback.

In a narrow sense, interaction tasks can be divided into basic interaction tasks, extended interaction tasks, ecological and scenario tasks, and system basic experiences.

The basic and extended interaction tasks are distributed in multiple function modules in the automotive HMI system, including media and entertainment, climate control, telephone, map navigation, and vehicle control. Common basic and extended interaction tasks are listed in Table 3.1.

Table 3.1 Common basic and extended interaction tasks

Module	Basic interaction tasks	Extended interaction tasks
Media and entertainment	Search for songs, volume control, next track, switch to a Bluetooth audio source	Play media, add to favorites, play the radio, play videos
Map navigation	Enter a destination and start the navigation, view full route overview, turn off voice broadcast, add waypoints, add gas stations/charging piles along the route, end navigation	Frequently used destination, check points of interest, adjust to detailed broadcast, access traffic information, switch to 2D map
Telephone	Enter a number and call it, dial a designated contact's number, answer a call	
Climate control	Turn off the climate control system, raise the temperature, fan speed control, switch to external circulation	
Vehicle control		Turn on seat heating, turn on defogging, adjust airflow to the head position, unlock the doors, open the trunk, turn on high-beam headlights, switch driving modes, open the left front window, open the right rear window, open the sunroof, open 50% of the sunroof, shut the sunshade, turn off the central information display, switch the ambient light color

Basic interaction tasks are used more frequently than extended interaction tasks and are more often performed during driving; thus, they are evaluated as secondary tasks. The test process of secondary tasks requires real road driving or simulated driving, and the evaluation of secondary tasks also needs to fully consider their impact on driving safety. Conversely, extended interaction tasks have a higher percentage of operations while the vehicle is stationary. When limited by the test time and environment, extended interaction tasks may not be evaluated as secondary tasks; thus, testing need not be performed under the driving environment.

Note that no strict demarcation exists between basic and extended interaction tasks; they are distinguished mainly for the convenience of the test process execution, which does not mean that extended interaction tasks can only be performed while the vehicle is stationary. If the test time and environment permit, full evaluation of all extended interaction tasks under the driving environment is possible, similar to the basic interaction tasks. Nevertheless, owing to their low utilization frequency during driving, they should be assigned a low weight in the evaluation even if they are tested in the driving environment.

During testing and evaluation, the initial state of basic and extended interaction tasks usually requires the central information display and the instrument cluster display to stay on the home page and the voice control to be inactive for the conformity of the interaction process. For other navigation tasks during the navigation guiding process, such as checking the route overview, turning off the voice broadcast, and adding waypoints, the initial state is selected as the navigation guidance page and the navigation guidance is required to be in progress. This is because during the navigation guidance process, most users usually stay on the navigation page, instead of returning to the home page, and do not stay on the pages of other functions for a long time.

Ecological and scenario tasks include information (e.g., news and weather), car service (e.g., maintenance and car wash appointments), life service (e.g., restaurant search and movie ticket booking), and social communication (e.g., in-vehicle WeChat and group travel). There is no apparent boundary between ecological and scenario tasks, and they are gaining popularity and expanding on an increasing number of models. These tasks are achieved when the vehicle is networked with the outside world, while some of them also require differentiated services based on the real-time location of the vehicle. Unlike basic and extended interaction tasks, ecological and scenario tasks, which have a relatively low utilization frequency and can be performed on smartphone applications, are usually not essential for automotive HMI systems. Therefore, these functions need to be sufficiently appealing to attract users.

The popularity of ecological and scenario tasks in vehicles is relatively low. Therefore, when a model is evaluated, whether it has some specific functions must be determined. This step is usually not necessary for basic interaction tasks, as all basic interaction tasks are currently available in mainstream intelligent vehicles. Owing to the diversity and low utilization frequency of ecological and scenario tasks, the interaction modality of each task need not be considered specifically to simplify the evaluation process.

The basic experience of the system, such as turning a page with the finger to see if the animation is smooth, or going to the music or navigation homepage to see if the layout is feasible and the design is pleasing, is a relatively free yet important experience for users. Although the user does not need to achieve a very specific purpose through a certain interaction task, it establishes the first impression of an automotive HMI system for the user. Especially in the process of purchasing a vehicle, many users are not likely to try multiple specific interaction tasks, but simply judge the merits and demerits of the system through a free experience.

In addition, interaction tasks in the broad sense include information reading and navigation performance. Information reading tasks, with a high occurrence frequency, are those tasks where the output can be read directly without the need for the user to input any additional information to the automotive HMI system, such as reading the vehicle and engine speeds on the instrument cluster or reading the orientation arrow from the navigation page. On highways, drivers read the speed every tens of seconds to make sure they are not speeding. On complex urban roads, drivers need to read the navigation information several times before each intersection to make sure they are in the correct lane.

However, sometimes information reading tasks are not the main object of automotive HMI system evaluation. For one thing, these tasks involve only visual and cognitive distractions rather than physical movements, which makes them more difficult to test and require specialized and expensive equipment. For another, such tasks are too simple. As long as there are no obvious design problems with the size and position of the information display, the performance of each vehicle is similar.

The navigation performance input does not belong to HMI, but the position, speed, and acceleration of the vehicle itself do. An excellent navigation system is expected to be able to determine the vehicle state quickly and accurately and provide smooth and precise route guidance.

3.2.2 Interaction Modalities

In the field of HMI, interaction modality is an independent type of sensory channel that is used when there is input/output between humans and machines [1]. In automotive HMI, the interaction modality is defined more specifically and is usually more focused on the input of information. The four input-oriented interaction modalities that are currently in widespread use in vehicles are central console buttons, central information displays, steering wheel buttons, and voice control [2].

(a) Central console buttons: The driver inputs information by pressing a physical button or turning a physical knob on the central console and receives feedback through the stiffness and damping of the buttons and knobs themselves. Additionally, feedback and output may take the form of lighting around the buttons and knobs, as well as changes in the information displayed on the central information display.

(b) Central information displays (abbreviated as CID, or central touchscreen): The driver enters commands by tapping and scrolling on the central information display and sees the output feedback on the display. Additionally, feedback may be conveyed via screen vibrations and audio ticks.
(c) Steering wheel buttons: The driver inputs information by pressing a physical button on the steering wheel spokes and receives feedback through the stiffness and damping of the button itself. Additionally, feedback and output may be delivered through changes in the information displayed on the central information display or the instrument cluster display.
(d) Voice control: The driver controls the HMI system through voice commands and receives feedback through the voice sent by the system as well as the information on the central information display.

As the oldest interaction modality of automotive HMI, central console buttons have been the most basic and common inputting method since the introduction of radios in vehicles in the 1920s. Steering wheel buttons, which are an extension of buttons and knobs, first appeared in the 1960s. Central information displays were first introduced in the 1980s; however, touchable displays began to gain mass popularity in the 2000s. The development history of voice interaction is even shorter, with truly practical systems being available for less than a decade.

Owing to the remarkable differences in their operating mechanisms, central console buttons and steering wheel buttons are regarded as two different modes in terms of interaction modality. The former is an arm movement where the driver needs to take the right hand off the steering wheel and use the whole arm to extend the hand to the central console, whereas the latter is a finger movement, where the driver's palm remains on the steering wheel while operating them. These two different operating mechanisms can affect the degree of driver distraction. Meanwhile, from the perspective of development history, the origin and popularity of steering wheel buttons lag those of central console buttons.

Among central-console and steering-wheel buttons, there is a special category called directional buttons, which can be one button or a set of up and down, left and right buttons, a knob, or a touchpad. Directional buttons, which have no fixed function and are typically used to select entries on the on-screen menu, must work together with the content on the central information display or the instrument cluster display. With the increasing popularity of touchable central information displays, there are fewer and fewer directional buttons in the central console. For example, in 2018, the Audi A8 removed the MMI knob (Fig. 3.2) and the Mercedes-Benz A-Class removed the COMAND touchpad. However, because the instrument cluster display is positioned behind the steering wheel and can hardly be touched, directional buttons are still important steering wheel buttons.

In recent years, touch-sensitive buttons, which, similar to induction cooker keys, have no press stroke, have also appeared in the central console of many vehicles. Some touch-sensitive buttons not only allow tapping but also support sliding gestures on their surface. Some touch-sensitive buttons have vibrating motors behind them that produce feedback similar to cell phone touchscreens. Touch-sensitive buttons

Fig. 3.2 MMI knob of the Audi A8 (2011 model), which was later removed in the new generation (*Source* Audi AG)

visually resemble the central information display, but they are still categorized as buttons rather than screens. The content displayed on the screen must be dynamic and variable, which allows different icons to be displayed in the same position at different times. Whereas the icons in the backlight of touch-sensitive buttons are fixed or with two or three backlight icons switchable with each other, which is why they are not categorized as screens.

The classification of interaction modalities is the most practical aspect of the evaluation process of automotive products. If the classification is broader, for example, if both the central console buttons and steering wheel buttons are regarded as buttons, there will be two different operating modes for the same task using the same interaction modality, which is difficult to distinguish. For example, the task of adjusting the volume on many models can be performed using either a central console button or a steering wheel button. If the classification is more elaborate, for example, if the central console buttons are classified into mechanical buttons, touch buttons, or knobs, there will be diverse interaction modalities in different models, making it difficult to compare them.

In addition to the four common interaction modalities discussed above, there are some emerging interaction modalities that may appear in vehicles in the future, such as in-air gesture control, eye-movement control, EEG control, and expression control. However, these interaction modalities are not yet widespread or can only be performed for a very limited number of interaction tasks; therefore, they will not be discussed in depth here as typical interaction modalities in the evaluation system.

Not all interaction tasks necessarily involve the four input-oriented interaction modalities discussed above. Some interaction tasks involve only output, such as observing the vehicle speed on the instrument cluster, observing navigation directions on the HUD, and listening to the system broadcast for navigation directions. Output-only interaction tasks are usually information-reading tasks in the interaction task classification, which are not usually used as the primary object of automotive HMI evaluation.

There are also some interaction tasks that involve input but are not directly operated by a human. For example, the climate control automatically adjusts the fan speed according to the temperature inside the vehicle. Another example is the navigation system recalculating the route when the car yaws in the wrong direction. Some of these tasks have a more complex trigger logic or require extreme conditions in the external environment to trigger, and therefore are less reproducible. For such tasks, making arbitrary conclusions during the evaluation process should be avoided, as some functional failures may simply be due to inadequate simulation of the test environment itself, rather than a problem with the product.

3.2.3 Evaluation Indexes

Evaluation indexes, which are independent of the interaction task and modality, are quantifiable evaluation dimensions in automotive HMI. The meaning of an evaluation index does not vary with changes in the interaction tasks or modalities, whereas some interaction tasks may not be applicable to some evaluation indexes. For example, some very complex vehicle system setup tasks, as well as video-playing tasks, are not expected to be operated while driving; thus, safety-related evaluation indexes are not applicable to such tasks.

A set of well-established evaluation indexes should be universal. The evaluation index itself is independent of the interaction task and modality, but it must also be adaptive to different interaction tasks and modalities. Additionally, an evaluation index should not have different interpretations for different interaction tasks, nor should it have different evaluation criteria for different interaction modalities. Furthermore, an evaluation index does not necessarily apply to all interaction tasks and modalities, but each evaluation index should cover as many tasks and modalities as possible. Otherwise, if an evaluation index is only applicable to one interaction task or one interaction modality, this part of the evaluation system is actually compressed into one or two dimensions, and the three-dimensional orthogonal structure loses its meaning.

For example, the lane-keeping index in safety applies to all tasks that need to be performed during driving simulations and to all interaction modalities. For central touchscreen tasks, this index is obtained by conducting tests that show complex tasks usually cause a more noticeable lane departure than simple tasks. Furthermore, for the same interaction task, the lane departure resulting from using the central touchscreen is often more prominent than that resulting from using steering wheel buttons as well.

When scoring the evaluation index test results, the same scoring criteria are generally used for different modalities of the same task. This is because the goals of all operations are identical and comparable and the same scoring criteria make it easy to determine which interaction modalities are poorly designed and need improvement. However, the scoring criteria for the same evaluation index may vary with different interaction tasks. For example, the task of inputting the navigation task is definitely more difficult than answering a call and would lead to increased driver distraction; thus, it would be unfair to use the same scoring criteria for both tasks.

A set of effective evaluation indexes should also have stability. Automotive HMI technology is developing rapidly, with new interaction designs and new hardware devices entering the market every year; however, the evaluation indexes cannot change frequently. Otherwise, it will be impossible to compare the evaluation results over time, which makes it difficult for car designers to predict product performance in future evaluations. This requires that the evaluation indexes be relatively abstract and not bound to specific technologies.

For instance, the index operation steps in efficiency are relatively abstract and is applicable to all information input-related interaction tasks, as well as to all interaction modalities. If this index is changed to "click steps" under the technical conditions of 2010, it is essentially equivalent to "operation steps" as it can be applied to almost all buttons and all tasks on the central information display at that time. However, over the past decade, touch technology on the central information display has shifted from resistive to capacitive, allowing more finger-swiping interactions on the screen than just "tapping." Moreover, voice control has spread rapidly in vehicles in the recent years, which cannot be measured by "click."

Although the evaluation indexes are stable, the scoring metrics for specific test scores can vary. For example, in a given year, the task time to enter the navigation destination in less than 8 s can be given a full score; however, with the continuous improvement of vehicle models on the market, the requirement for full score may be reduced to 6 s within several years.

Evaluation indexes, as the most important and complex component of the three-dimensional orthogonal evaluation system, will be introduced and discussed in detail in subsequent chapters.

3.2.4 Evaluation Items

Evaluation items, as the basic unit in the automotive HMI test and evaluation process, are composed of the intersection of three dimensions: interaction tasks, interaction modalities, and evaluation indexes, among which the combinations of interaction tasks and modalities are called test cases. Evaluation items are units after dividing test cases according to evaluation indexes.

For example, lane keeping in the process of inputting a navigation destination via voice control is an evaluation item, as displayed in evaluation item A in Fig. 3.3, where the voice control is the interaction modality, inputting the navigation destination is the interaction task, and lane keeping is a third-level evaluation index under safety. In another example, the text size displayed on the screen when searching for a specific song with the central touchscreen is also an evaluation item, as shown in evaluation item B in Fig. 3.3, where touching the central information display is the interaction

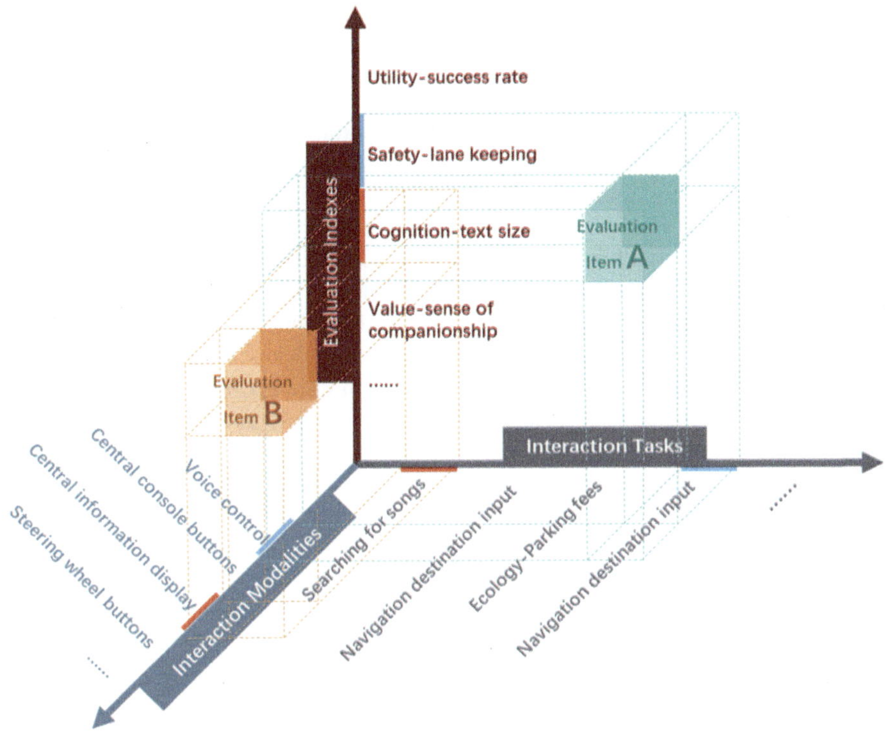

Fig. 3.3 Two typical evaluation items in the three-dimensional orthogonal evaluation system

modality, searching for songs is the interaction task, and the text size is a third-level evaluation index under cognition.

If 200 interaction tasks, four interaction modalities, and more than 40 evaluation indexes are present in an automotive HMI evaluation system, then the evaluation system would theoretically have more than 30,000 evaluation items, which poses serious difficulties for its practical implementation. However, there are some evaluation items that theoretically are not important to the test and evaluation or are meaningless themselves, which do not need to be included in the actual evaluation system. For example, all indexes included in safety apply only to the interaction tasks that are frequently used while driving rather than to rarely-used interaction tasks while driving. Another example is the tasks under voice interaction, which mainly rely on the invisible interface of voice with no displayed elements, while the visible interface of the central information display simply serves as an aid. Therefore, evaluation indexes such as understandability in cognition are meaningless for voice interaction. Nevertheless, a complete HMI evaluation system would have at least 2,000 evaluation items. If the evaluation is more detailed, this number can reach 4,000 or more.

3.3 Advantages of Three-Dimensional Orthogonal Evaluation Systems

Thousands of evaluation items are present in an automotive HMI evaluation system, and without a good structure, integrating all items in one system that is free of omission, non-overlapping, and expandable is difficult. Therefore, a three-dimensional orthogonal evaluation system is of vital importance for automotive HMI evaluation.

The common evaluation system seen in other fields is usually a tree structure consisting of first-, second- and third-level directories, and so on. Similar to a large tree, the trunk is divided into several main branches, which are further divided into side branches where leaves grow. The evaluation indexes introduced above also follow a typical tree structure, but each evaluation index is only one dimension of the entire three-dimensional orthogonal evaluation system and the entire evaluation system does not follow a tree structure.

If the entire evaluation index follows a tree structure, the establishment of this structure will encounter the following issue: should the total score be divided into the scores of each interaction modality first and then into the scores of each interaction task, or should the total score be divided into the scores of each interaction task first and then into the scores of each interaction modality? If we choose the former, we will not be able to answer the question of how well a vehicle's climate control tasks perform because they are scattered across the interaction modalities. If we choose the latter, we cannot answer questions about how well a vehicle's voice interaction performs because they are spread out among interaction tasks.

The symmetry of the three dimensions in a three-dimensional orthogonal system can address this problem very well. We do not need to think about which interaction modality and task needs to be ranked higher because they are all one of the three orthogonal dimensions, which has no hierarchical ranking.

We can flexibly choose a "section" in the three-dimensional orthogonal system to be explained as needed. When we need to judge the performance of a vehicle's climate control task, we can take the interaction task as the section and select all the tasks related to climate control for analysis. When we need to judge the performance of a vehicle's voice control, we can take the interaction modality as the section and select all the voice control tasks for analysis. When we need to know the efficiency performance of a vehicle, we can take the evaluation index as the section and select all the evaluation indexes under the efficiency for analysis.

In addition, the three-dimensional orthogonal system is also very conducive to the expansion of the scope of the evaluation system. The extension of one dimension does not affect the other two dimensions. If we were to add autopilot-related operations to the evaluation system, we only need to extend the dimension of the interaction tasks rather than the other two dimensions of the interaction modalities and evaluation indexes. If, some years later, expression control becomes popular in automotive HMI, then only the dimension of the interaction modalities needs to be extended without changing the other two dimensions of the interaction tasks and evaluation indexes.

If the automotive HMI evaluation system does not adopt a three-dimensional orthogonal system but uses a tree structure, then the first-level directory in this structure is likely to include basic tasks (or common tasks), map navigation, and voice control, among other aspects, which may seem intuitive and is consistent with the tester's experience flow. However, this structure has problems. First, map navigation is a set of tasks, whereas voice control is an interaction modality. The two are not in the same dimension and are not juxtaposed. A simple example is whether inputting a navigation destination using voice control should belong to the map navigation directory or the voice control directory. Actually, it belongs to both. There is overlap between these two first-level directories. Second, there will be a logical dilemma in developing such an evaluation system. Let us say that expression control becomes prevalent in automotive HMI some years later, with which people can perform simple tasks such as controlling the climate or music. Then, should the test items corresponding to expression control belong to the basic tasks, or should a new separate first-level directory be created for it? If incorporated into the basic tasks, expression control cannot be parallel to voice control. If a new separate first-level directory is created, it will be too easy for the first-level directory of the entire system to change, and the total score of the new and old test results will be different and, therefore, are not comparable. As can be seen, a tree-structured evaluation system has some insurmountable problems, such as logic overlap or difficult expansion, emphasizing the need of developing the three-dimensional orthogonal system.

References

1. Karray F, Alemzadeh M, Saleh J A, Arab M N. Human-computer interaction: overview on state of the art [J]. International Journal on Smart Sensing and Intelligent Systems, 2008, 1(1).
2. Ma J, Gong Z, Tan J, Zhang Q, Zuo Y. Assessing the driving distraction effect of vehicle HMI displays using data mining techniques [J]. Transportation Research Part F: Traffic Psychology and Behaviour, 2020, 69: 235–250.

Chapter 4
Formulation of Automotive HMI Evaluation Indexes

4.1 Design Objectives for Automotive HMI

Evaluation indexes play a crucial role in the three-dimensional orthogonal automotive HMI evaluation system. Accordingly, when evaluation indexes are formulated, the design objectives of automotive HMI systems should be considered. Automotive HMI systems are not simple engineering systems. They have complex design objectives, with rational and emotional aspects, as well as tangible and intangible parts. The design objectives of automotive HMI can be divided into three levels from low to high, namely, functionality, usability, and imaginability, as shown in Fig. 4.1.

4.1.1 Functionality

Functionality refers to the software and hardware of the automotive HMI system, and the functions that can be achieved. In general, a more superior automotive HMI system is the one equipped with more software and hardware, and can achieve more functions.

The term "functionality" corresponds not only to interaction tasks but also to interaction modalities. For instance, all vehicles with navigation systems can input navigation destinations, but compared to touch inputs on the central display, inputting destinations via voice control and sending destinations through mobile applications are more innovative functions. Occasionally, functionality places even greater emphasis on interaction modalities than on interaction tasks. For example, we can say that the heads-up display (HUD) is a relatively innovative function, but when describing it, we are less concerned about what can be displayed in this HUD system than in the novelty of the system itself.

In general, the functionality of a car can be described by the vehicle model configuration table. For example, Fig. 4.2 shows some of the configurations related to the

J. Ma and Z. Gong, *Automotive Human-Machine Interaction (HMI) Evaluation Method*, Research on Automotive Intelligent Cockpit,
https://doi.org/10.1007/978-981-99-9951-4_4

Fig. 4.1 Three levels of the design objectives of automotive HMI

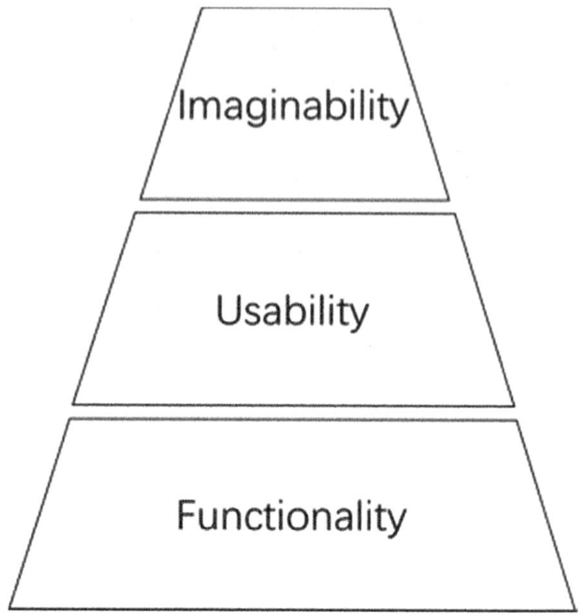

HMI system taken from the vehicle model configuration table of the Mercedes-Benz S-Class 2023. Among them, the digital radio is predominantly an application software included in the HMI system; the 3D driver display is primarily a set of hardware; the pre-installed navigation services are mainly reflected as output information, which needs to run on the navigation software system. Thus, evidently, the content encompassed by a functionality is varied without a standardized pattern.

Functionality is sufficient when describing some very simple functions, as in the case of a light switch in a vehicle, which generally has only two states: on and off. Regardless of how the switch is designed, the user experience (abbreviated UX or UE) is typically similar as long as there are no basic errors.

However, functionality only describes whether a certain function is available, not how effectively it works. Owing to the complexity of automotive HMI systems, functionality is limited when used to describe such systems. For example, all vehicles with a navigation system are capable of entering a navigation destination. If we consider the typing words task of the destination inputting process as one step, then the navigation systems of some vehicles can complete the task with only four steps, whereas other vehicles require as many as nine steps. Evidently, a four-step process is more convenient and works better than a nine-step process.

In the current technological environment, all functional controls in the cockpit, except for those that are highly relevant to driving (e.g., light switch and driving mode selection), can only be achieved through a set of operating procedures rather than in one step with a physical button. As there is a set of procedures, their usage experience cannot be clearly described by determining whether a certain function is present. Therefore, we need to introduce usability for a more in-depth analysis.

Standard equipment S-Class

● Standard – Not available

Communications and In-Car Entertainment	AMG Line Premium	L AMG Line Premium	L AMG Line Premium Plus	L AMG Line Premium Plus Executive
3D driver display - 12.3-inch freestanding display	●	●	●	●
5G communications module[2]	●	●	●	●
Accident recovery and breakdown management[2,3] - in an emergency quickly provides a connection to the Customer Assistance Centre in order to obtain assistance, for example with recovery of the vehicle or from the breakdown service Mercedes-Benz Service24h	●	●	●	●
DAB digital radio	●	●	●	●
Fingerprint scanner - colour-illuminated fingerprint scanner in the trim strip below the OLED media display, two fingerprints per user can be stored, quick access replaces PIN entry, protects personal data such as profiles plus office and payment functions	●	●	●	●
Hard-disc navigation	●	●	●	●
Geofencing[2] - available to use free of charge for three years following activation[3], denotes a specific area and notifies via push message when the vehicle leaves or enters this area	●	●	●	●
Maintenance management[3] - proactively transfers maintenance details to the preferred Mercedes-Benz service partner in order to coordinate workshop appointments	●	●	●	●
MBUX augmented reality head-up display	–	–	●	●
MBUX augmented reality for navigation	●	●	●	●
MBUX extended functions[2]	●	●	●	●
MBUX Interior Assist (front)	–	–	●	–
MBUX Interior Assist (front and rear)	–	–	–	●
MBUX navigation premium	●	●	●	●
MBUX rear tablet	–	–	–	●
OLED central display - 12.8-inch touchscreen	●	●	●	●
Parked vehicle locator[1] - available to use free of charge for three years following activation[2], shows the position of the parked vehicle in a radius of approx. 1.5 km	●	●	●	●
Personalisation[1] - available to use free of charge for three years following activation[2] stores settings such as the radio station or navigation destination in the personal driver profile and enables these to be used in vehicles with the same telematics generation, the profile can be linked to the key, and the functions can therefore already be activated before getting in	●	●	●	●
Pre-installation for Car Sharing[1]	●	●	●	●
Pre-installation for Navigation Services[1] - display of online content on the navigation map, parking spaces, weather information, local search	●	●	●	●
Remote and Charging Services Plus[1] - available to use free of charge for three years following activation[2] (S 580 e only)	–	●	●	●
Remote Door Locking and Unlocking[1] - available to use free of charge for three years following activation[2], provides convenient remote monitoring	●	●	●	●
Remote parking package[1] - available to use free of charge for 3 years following activation[2]	●	●	●	●
Remote retrieval of vehicle status[1,2] - shows mileage, tyre pressure, washer fluid level, etc.	●	●	●	●

Fig. 4.2 Excerpt of the vehicle model configuration table of the Mercedes-Benz S-Class 2023 (*Source* Mercedes-Benz official website)

The functionality of automotive HMI systems can sometimes limit usability. The addition of functions results in more distractions for drivers when searching for a specific function, which increases the difficulty of the task process. The conflict between functionality and usability is particularly pronounced for physical buttons. For example, the Porsche Macan has more than 30 functions laid out on the central console in the form of physical buttons (Fig. 4.3), which are difficult to remember and operate without looking. Accordingly, more effort is required from the user

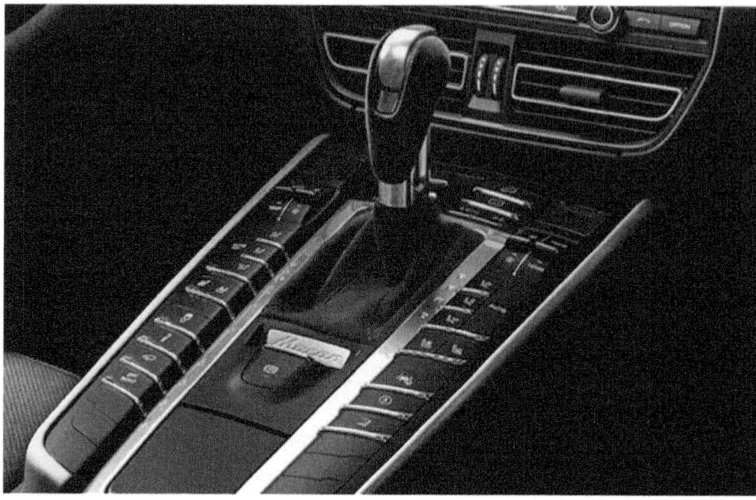

Fig. 4.3 Central console of the Porsche Macan (2018) (*Source* Porsche AG)

to locate a specific button. Interactions on the central information display can alleviate the conflict between functionality and usability through appropriate information structure design; however, this conflict cannot be completely eliminated.

4.1.2 Usability

Usability is a new concept involving the interactions among users, products, tasks, and environments. Many researchers believe that there is no authoritative and perfect definition for usability. According to the international standard ISO-9241, the usability of a product is the effectiveness, efficiency, and satisfaction perceived by specified users when using a product in specified scenarios to achieve specified goals [1]. This description can be considered to define the research scope of usability rather than defining its essence and connotation.

Occasionally, the concepts of usability and user experience are interchangeable. The international standard ISO-9241 describes user experience as a user's perceptions and responses when using a product, system or service. This definition can be considered as user experience in a narrow sense. Here, the scope of user experience is basically the same as that of usability but is more inclined toward the subjective feelings of users.

However, since the popularization of smartphones, the meaning of the term user experience has begun to expand into various contexts. user experience represents not only "the experiences of users" but also a set of procedures and interface design in electronic products, a stage in the product development process, as well as a product design value. Thus, not only can we say "This phone has a great user experience" (referring to human perceptions and responses) but we can also say "This phone has a completely new user experience" (referring to the product design), "someone works in the user experience department" (referring to the research and development stage), and "they developed a product design approach based on user experience" (referring to ideas and values). It is precisely due to the confusion in the usage of the term UX in real contexts that we have tried to minimize its use and adopted the term "usability" instead when conducting research with rigorous boundaries.

The term "usability" first appeared in the 1930s. On March 8, 1936, General Motors (not General Electric) introduced the concept of usability for the first time in a refrigerator advertisement in The Palm Beach Post (Fig. 4.4). When explaining usability, the ad mentioned that the refrigerator was handier to use, saves steps, and saves work, which are the three aspects of usability that are still applicable today [2, 3]. For example, a laptop with a screen that can be opened with one hand is easier to use than one that requires two hands. There are handy drop-down menus on a cell phone interface to enable or disable frequently used settings, which reduces the number of steps compared to a lengthy settings menu. A vehicle with an electric trunk lid can help the user save energy while opening and closing. However, early definitions of usability did not address the relatively subjective aspect of satisfaction, which was partly due to the immaturity of relevant research, and partly because household appliances at the time were simple to operate, which did not require extensive discussions on subjective satisfaction.

The usability of automotive HMI consists of another crucial component: safety. Drivers use the HMI system while driving instead of giving it their full attention. Driving the vehicle itself is the primary task, whereas using the HMI system is a secondary task. However, a driver's attention span is limited. If the driver attempts to perform any secondary task, the distraction may lead to a decline in driving performance. Other secondary tasks include talking to other passengers, glancing at the scenery outside the vehicle, talking on a cell phone, eating, smoking, and so on. However, these secondary tasks are beyond the scope of this book.

Why is safety part of usability rather than functionality? As a secondary task, using the automotive HMI system can indeed cause driver distraction and affect driving safety, but the scope of its influence is usually limited. If no major design problems exist, then it is unlikely that a function that should be used while driving cannot be performed at all due to a safety hazard.

Theoretically, a situation may occur where a function cannot be used at all during driving due to potential driving safety risks. For example, in a sedan with two rows of seats, if the temperature adjustment interface of the rear independent climate control system is located only at the rear end of the central armrest instead of the central console in front of the driver, the driver must twist their body significantly to adjust the rear temperature. This operation can be safely performed when the vehicle is stationary; however, in a moving vehicle, it poses a serious driving safety risk and should be avoided. In this situation, safety affects functionality. However, such extreme examples are very rare and typically make little sense. In this case, why would the driver want to adjust the rear temperature while driving instead of

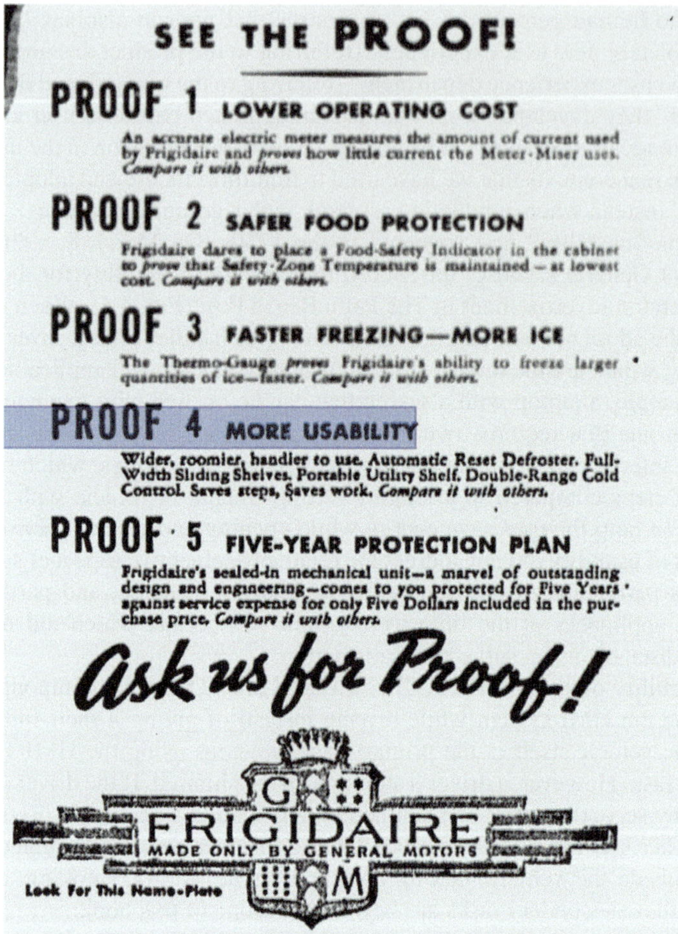

Fig. 4.4 Advertisement for a General Motors refrigerator mentioning "usability" (*Source* The Palm Beach Post)

allowing the passengers in the rear row to adjust it themselves? This is a function that lacks practical significance.

4.1.3 Imaginability

Imaginability refers to what users can imagine when using or planning to use a product beyond functionality and usability. In previous academic research, imaginability generally did not refer to a strict research scope or research direction, but was only a common academic term. Nevertheless, imaginability is critical for automotive HMI studies because functionality and usability are not sufficient to explain the entirety of the user's perceptions when using a product. Therefore, we must use imaginability to explain other more subjective factors.

In the real market, even a car (not just an HMI system) that possesses both excellent functionality and usability may not necessarily be successful, which is a phenomenon that is particularly obvious in today's Chinese automotive market. Before 2010, most car brands (both international and local) were able to achieve relatively optimal market performance because most automotive products did not have evident shortcomings in terms of functionality and usability. However, since 2015, some French, American, and Chinese manufacturers have seen a dramatic decline in sales, with many of them experiencing more than 50% downswing compared to their peak. This was not caused by a regression in functionality and usability but because Chinese car consumers had a growing demand for product imaginability, which these manufacturers failed to meet.

Compared with the overall performance of automotive products, does imaginability account for a lower or higher proportion in the automotive HMI system? Automotive HMI systems have a wide variety of functions and designs, with considerable differences among products. Therefore, theoretically, if users can fully understand the HMI system of each vehicle, then functionality and usability will play a larger role in their judgment of the merits and demerits of the product, which also implies that imaginability will account for a relatively small proportion. Nonetheless, owing precisely to the diverse functions of automotive HMI systems, it is almost impossible for users to have a full and sound understanding of a vehicle's HMI system before buying the vehicle. In fact, many users use only some of the functions in the HMI system even after several years of owning the vehicle. Therefore, whether during the purchase or usage process, the user's judgment of the HMI system is to some extent one-sided and irrational, which renders imaginability even more important.

The word "imagination" is a common term in daily life. However, in many disciplines, particularly in engineering disciplines, it is relatively unfamiliar. Therefore, we find it necessary to explain the importance of imagination to human organization and behavior. In his 2012 book *Sapiens: A Brief History of Humankind,* Yuval Noah Harari, an Israeli historian, divided the evolutionary history of *Homo Sapiens* from the Stone Age to 21st Century into four major stages. The first of these was referred to as the cognitive revolution (approximately up to 70,000 BC), during which imagination evolved in *Homo sapiens* and a language capable of describing stories emerged. The book also mentions that all large-scale human cooperation is rooted in some type of fictional stories that exist only in the collective imagination. Thus, gods, nations, money, laws, corporations and brands exist and function only in common human imagination [4]. In 1983, Benedict Anderson, a political scientist and historian, published a book titled *Imagined Communities: Reflections on the Origins and Spread of Nationalism,* stating that nations are imagined communities of people. In history, nationalism emerged before the formation of nations, not the other way around [5]. Evidently, imagination plays a crucial role in human development.

Product imaginability usually arises from three aspects, namely, brand, product technology, and product design.

A brand is a name, term, design, symbol, or other characteristics that distinguish a seller's product or service from others [6]. In everyday use, the most prominent form of a brand is the trademark, including word and figurative marks. For example, Mercedes-Benz (Mercedes for short) is one of the most valuable automotive brands in the world, and its figurative trademark, a "three-pointed star", has almost the same impact as the brand name (i.e., its word mark). By contrast, Daimler AG, to which the brand belongs, has less public influence than the brand itself.

Compared with the product itself, the brand has stronger stability. The quality of the products offered by a brand may vary, but this does not necessarily lead to drastic fluctuations in the value of the brand to the user. In fields such as household appliances and light commercial vehicles, some products that are almost identical can be sold at different prices under different trademarks. The brand itself does not represent the actual product performance, but the imagination it evokes in the user. For instance, a pair of Adidas sneakers will make the user feel more energetic. If the brand collaborates with another trendy brand, the user of the sneakers will not only perceive themselves as more energetic but also fashionable. These imaginations or perceptions are not necessarily related to whether the user actually likes to exercise.

The imaginability of an automotive brand is crucial. Furthermore, due to the high technical complexity of automotive products, the image that the brands present to users usually places more emphasis on technical capability. These imaginations may be categorized into two types.

The first type is the brand background and narrative. For example, Mercedes-Benz is the inventor of automobiles; Aston Martin is James Bond's car; the founder of Tesla is striving to launch rockets into space. Such imaginations have virtually no association with product capability in itself but are still vital. Although difficult to portray, because the brand needs an extremely powerful narrative to be effective, it is remarkably stable because this abstract perception, divorced from the product, is less susceptible to fluctuations in the performance level of the product itself.

The second type is the technology and capabilities that the brand has mastered. For example, Audi vigorously promoted turbocharging and full-time all-wheel-drive technology in the 1980s, which not only achieved outstanding results on the race track but also established Audi's image as a technological pioneer and a genuine luxury brand. Even if the user buys a naturally aspirated, front-wheel-drive Audi model, they will still feel closer to the most advanced automotive technology of their time. The ripples of technology can affect the brand. This is why many people choose the lower end of a luxury brand over the higher end of a mainstream brand.

In addition to brand, product technology can also enhance the user's imagination. The primary goal of product technology is, naturally, to achieve product performance. However, for many products, technology is either excessive or not readily perceptible to most users. Users actually own these technologies and features, but they have little practical value. For example, a vast majority of car users cannot tell the difference in driving between six- and four-cylinder engines, rear- and front-wheel drive, or multi-link and trailing-arm suspension. However, consumers still pay for theoretically superior technology. Even the acceleration performance of a Ferrari or the off-road

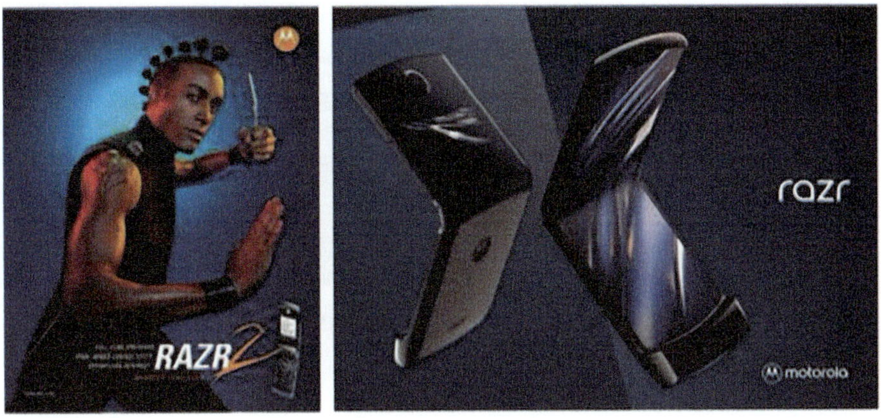

Fig. 4.5 Advertisements for the Motorola RAZR2 V8 (2007) and RAZR (2019) (*Source* Motorola, Inc.)

capabilities of a Mercedes G-Class are merely figments of imagination for some car owners because they never use them in their daily driving.

In recent years, users have been gradually paying more attention to the technical features of the products themselves. This trend was largely led by the cell phone industry. Before the prevalence of smartphones, cell phone advertisements were usually created with human characters as the main element, often regarded as more important than the phone itself. In contrast, given the current popularity of smartphones, most cell phone advertisements portray the product itself as the sole protagonist and highlight features related to new technologies. On the official websites of cell phone companies, the contents of product introduction usually present only technical parameters and pictures or videos related to the technology, instead of emphasizing the feelings of the characters and the ambience of use as it did over a decade ago. As can be seen in Fig. 4.5, the advertising styles of the same series, the Motorola RAZR, are completely different in the two eras.

Similarly, the technical parameters and advanced functions of products play a more important role in the promotion and publicity of automotive products. For example, in Fig. 4.6, many technical parameters are mentioned when explaining the aerodynamic design and in-vehicle camera system of the IM L7. Most users do not necessarily have the ability to objectively judge whether these technical parameters can bring about real enhancements in experience, but the process of reading such parameters itself builds the users' imagination.

Product design is the third source of imaginability. The design of a product can directly incorporate what it expects the users to imagine by depicting it as a pattern or model, which is a simple practice. For example, the shark head design on the nose cone of the Jordan EJ11, a Formula 1 racing car, can conjure up images of the car being as fierce as a shark, as shown in Fig. 4.7. Product design can also convey macro and abstract imaginary concepts to users. For instance, the first-generation Lexus NX

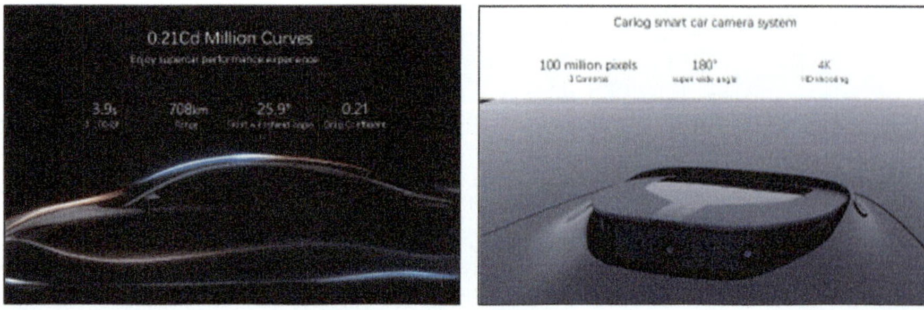

Fig. 4.6 Product introduction of the IM L7 (2021) (*Source* IM Motors official website, translated from Chinese language)

and Range Rover Evoque both had futuristic shapes when they debuted. It was as if users could be transported to the future when choosing these products.

The imagination that comes from product design can also meet the user's spiritual and value demands. Rather than invoking a specific imagery, these designs usually resonate with the user's inner demands, making the vehicle more than a simple, cold machine. A case in point is Nomi, the avatar in NIO vehicles, as shown in Fig. 4.8. Nomi is a spherical robot mounted above the instrument panel. Its front screen can display various expressions that simulate a human face. In addition, the robot as a whole can rotate and "look" at a specific person in the car when having a conversation. Nomi works in conjunction with the voice interaction system for a more anthropomorphic interaction. This design is extremely popular in the Chinese market; it has not only become an iconic label for the NIO brand but has also been imitated in various forms by domestic and international automakers.

Fig. 4.7 Jordan EJ11 Formula 1 racing car (2001) (*Source* Jordan Grand Prix)

Fig. 4.8 Nomi, the in-vehicle avatar in the NIO ES6 (2020) (*Source* NIO Inc.)

Nomi has no obvious functionality in itself. Even if the user chooses not to pay for the optional avatar, NIO's voice interaction system can still perform all conversational functions, as Nomi is only an auxiliary output of the voice interaction system. Moreover, it only makes a limited contribution to the usability of the HMI system. It neither improves the success rate nor does it reduce the steps involved in the tasks. Its contribution to usability is mainly to assist users in understanding the current status of the HMI system. For example, when the user awakens the voice interaction system, Nomi turns its head around to indicate that it is listening to the user's commands. When music is playing, Nomi's expression changes with the music to indicate that music is being played by the system. However, the role of Nomi as a cognitive aid is not irreplaceable, as it can be achieved through dynamic icons on the central information display or dynamic ambient lighting in the vehicle without the avatar. Nomi's value is mainly reflected on the level of imaginability. With the avatar, the vehicle is transformed from a machine to a living, emotional partner. Companionship is a very valuable demand for Chinese costumers, which is fulfilled by the imagination brought by Nomi.

Brand, product technology, and product design can all enhance product imaginability. However, for automotive HMI evaluation, the brand factor is not usually considered because it is inherent to a product and cannot be easily changed. If some of the HMI interfaces can effectively reinforce the genes of the automotive brand, then these can be considered as successful product designs rather than be considered as the success of the brand itself. Additionally, in the automotive HMI evaluation process, product technology is less important than product design. This is because the imagination invoked by product technology in users usually arises during product promotion, rather than product use. Consequently, there is limited prospect for automotive HMI to strengthen the user's image of product technology. Therefore, product design is the most important aspect to evaluate when examining the imaginability of automotive HMI systems.

4.2 Evaluation Index System

Functionality, usability, and imaginability are the design objectives of automotive HMI. However, it is difficult to directly evaluate these objectives. Hence, they need to be divided into several first-level indexes in an evaluation index system, to facilitate intuitive testing and evaluation, as shown in Fig. 4.9.

Some evaluation indexes are not limited to a single level of HMI design objectives. However, these indexes may have a greater emphasis on a particular design objective. Therefore, there is a relatively obvious one-to-one correspondence between design objectives and evaluation indexes. Functionality is mainly reflected in utility. Usability is primarily embodied by efficiency, safety, cognition, and intelligence. Imaginability is mainly represented by value and aesthetics.

These evaluation indexes can be further categorized into two groups: rational and emotional. Between them, rational indexes include utility, safety, and efficiency, whereas emotional indexes include cognition, intelligence, value, and aesthetics, as shown in Fig. 4.10.

4.2.1 Rational Indexes

Rational evaluation indexes regard humans and machines as a highly simplified system. The inputs and outputs of the system have a clear cause-and-effect relationship. When the system is given a task command, it produces a relatively definitive set of driving and interaction task performance results, which are influenced by the HMI within the system, as shown in Fig. 4.11. The measurement of rational indexes is more oriented toward the measurement of results, and the methods are more similar to engineering tests. For example, if an interaction task requires a more complex operation, then it must have a longer operation time, with the two exhibiting

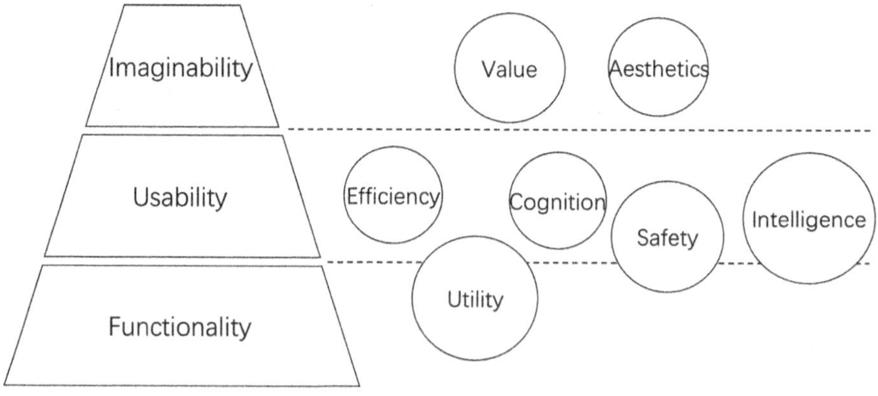

Fig. 4.9 Automotive HMI objectives and corresponding evaluation indexes

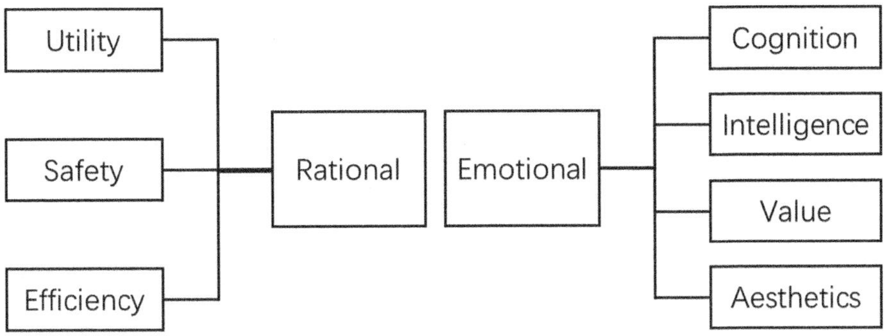

Fig. 4.10 Seven first-level indexes in automotive HMI evaluation

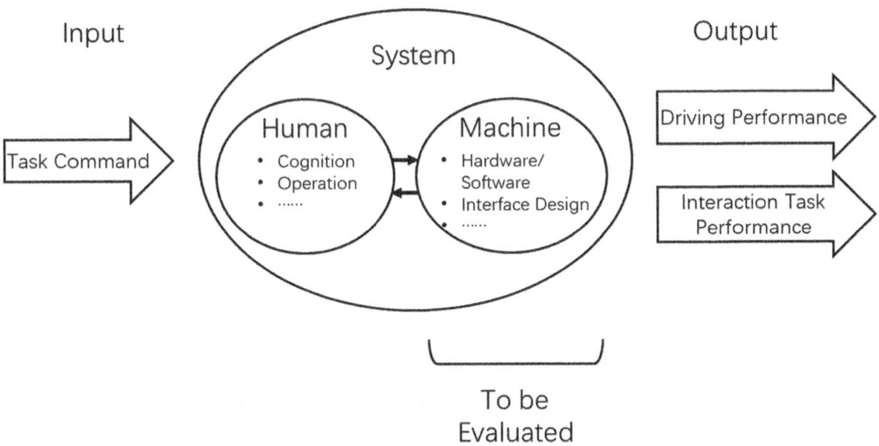

Fig. 4.11 Perspective of rational evaluation indexes

a strong causal relationship. Hence, we only need to measure the task time to infer the complexity of the task. Another example is, if the driver needs to tap an icon on the screen that is far away and which requires leaning over, then this icon will have a definite impact on the driver's lane-keeping ability during driving. There is also a strong causal relationship between the two. Thus, we can measure the degree of lane departure to determine the potential safety risk of using the HMI system while driving [7].

Rational indexes for automotive HMI evaluation are mainly based on the ISO-9241 international standard, in which usability covers effectiveness, efficiency, and satisfaction. Among them, effectiveness and efficiency are chosen as the basis of evaluation for the two first-level rational indexes. As ISO-9241 is not dedicated to automotive HMI, the first-level indexes of automotive HMI evaluation require further refinement to improve their relevance. In addition, satisfaction, which is more subjective and emotional, is not applicable as a rational index and will be elaborated further in the subsequent discussion on emotional indexes.

Utility refers to the ability of the automotive HMI to perform a specified interaction task effectively, accurately, and stably in a characteristic way. More specifically, utility encompasses the ability to achieve a specified interaction task through a specified interaction modality, the success rate of completing the task, and the ability to perform the interaction task consistently in a complex work environment. Utility is derived from effectiveness in ISO-9241, but covers a wider scope.

Efficiency refers to the relevant resources consumed by the user and HMI system when completing a specified interaction task. For the user, these resources mainly include the task time, physical movements, and visual attention. For the HMI system, they mainly involve the response time. Automotive HMI efficiency generally demands the interaction tasks to be viewed as secondary tasks, i.e., the driver is not fully focused on operating the HMI system, but rather prioritizing driving safety during the process of driving. Compared to electronic products such as smartphones, automotive HMI design faces a greater challenge when it comes to attaining higher efficiency.

Safety must also be included in automotive HMI evaluation, which is beyond the scope of ISO-9241. This is because, operating the HMI system while driving will inevitably lead to driver distraction, which may cause safety hazards. Safety refers to the ability of the automotive HMI system to reduce driving distractions and improve safety when driving the vehicle while simultaneously performing HMI tasks. It is unique to automotive HMI system evaluation because it does not evaluate the HMI system itself but rather the impact of it as a set of secondary tasks on a separate set of primary driving tasks. Such evaluation indexes are uncommon in other industries. For example, when evaluating a smartphone application, we do not consider the user's ability to watch TV while using the application. The balance between secondary and primary tasks is a major challenge for automotive HMI design and evaluation.

There is a sequential relationship among the three rational indexes. The functions in an automotive HMI system should first be functional, i.e., utility; second, they should be feasible while driving, i.e., safety; finally, they need to be handy, i.e., efficiency.

4.2.2 Emotional Indexes

Emotional evaluation indexes address the user's responses and feelings when using the HMI system. These indexes do not regard humans and vehicles as an integrated system but instead place more emphasis on the complexity of humans, encompassing different levels such as cognition, motivation, values, and aesthetics. These indexes aim to explore the mechanisms that affect the human responses and feelings, whereas the driving and interaction task performances are only indirect outcomes from the perspective of perceptual evaluation, and are not the focus of research in this area, as shown in Fig. 4.12. There are numerous emotional evaluation indexes, and diverse

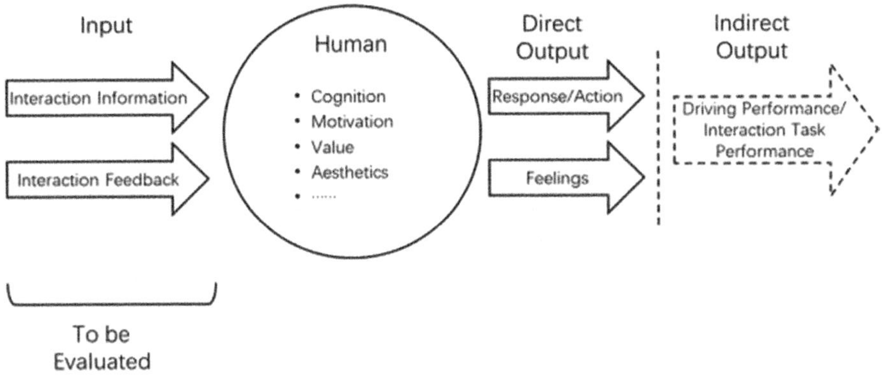

Fig. 4.12 Perspective of emotional evaluation indexes

methods to measure them. For instance, if there are too many elements in the central information display, it will be difficult for the user to find a specific element. An anthropomorphic avatar design can enhance the value of companionship to the user and thus create a greater reliance on the product.

Evaluating automotive HMI systems from a perceptual perspective is crucial yet challenging. Satisfaction, as defined in ISO-9241, is the extent to which the user's physical, cognitive, and emotional responses resulting from using a product or service meets the user's demands and expectations. This definition is more abstract than that for effectiveness and efficiency, which makes obtaining direct executable first-level emotional evaluation indexes more difficult. Therefore, these indexes must be set based on studies related to automotive-HMI and the three levels of automotive HMI design objectives (Fig. 4.1).

At the usability level, the cognition and intelligence indexes are needed. Cognition is the ability of automotive HMI to facilitate the user's perception, comprehension, memorization, and application in an accurate and efficient manner during use. Detailed indexes in cognition are derived from cognitive psychology, which is the most important foundational discipline in interaction design. Cognition is crucial for perceptual evaluation; in that it can offer a causal explanation for the usability level of a system.

Intelligence is the ability of automotive HMI to provide comprehensive, proactive, and accurate services to users in complex usage scenarios. In this context, the definition of intelligence is derived from the definition of human intelligence. Compared with the broader meaning of the word in the industrial context, intelligence here is more focused, emphasizing the expansion of the boundaries of HMI system capabilities. Higher intelligence can enhance functionality, and more importantly, it can continuously optimize usability to achieve a proactive and intuitive experience.

At the imaginability level, the two evaluation dimensions of value and aesthetics are essential. Value is the extent to which the user's imagination resulting from the automotive HMI matches with their existing values. In general, designs that rely on imagination to enhance value do not directly improve functionality or usability.

Value reflects the symbolic meaning of the HMI system to the user. The underlying foundation of value is culture, which in turn is the root cause for the distinctive needs of a particular user group. Therefore, cultural research is the cornerstone of value evaluation.

Aesthetics is the extent to which the visual interface design of an automotive HMI system conforms to typical aesthetic trends. It is the most intuitive feeling elicited by the interface design in the user or, in layman terms, it defines whether a design looks good or not. The dimensions of aesthetic index mainly consider the plane composition, color composition and three-dimensional composition, which together constitute the basis of art design.

Aesthetic evaluation is highly subjective; nevertheless, it should still be incorporated in the evaluation system because it is an important criterion by which many users subjectively evaluate an HMI system, making it an indispensable component. However, when selecting the objects for evaluation, we narrowed their scope to the extent possible by focusing only on the in-screen graphical interface design without considering the overall layout of the cockpit and styling of the physical buttons, as the former can be more easily abstracted into principles and recommendations.

4.3 Correlation and Positioning of Evaluation Indexes

4.3.1 Rational and Emotional Indexes

Rational and emotional indexes investigate the quality of the automotive HMI from different perspectives. However, they are also interrelated.

The evaluation results of cognition can sometimes be the cause for safety and efficiency issues. For example, excessively small text that is not readily legible is a visibility element in cognition, which is likely to cause a decrease in driving performance with respect to safety and an increase in task time with respect to efficiency. Although the relationships between cognition, safety, and efficiency are mutually supportive, they are not completely equivalent. Cognition has a broader scope, indicating that not all cognitive deficits will necessarily lead directly to safety or efficiency issues. Conversely, safety and efficiency performances are the result of the overall HMI design, which cannot necessarily be partitioned and traced to one or two specific cognitive indexes. Therefore, cognition, safety and efficiency are all integral.

The intelligence and utility evaluation indexes may be to some extent similar in their evaluation format for certain functions. For example, they both examine whether the HMI supports a specific function. However, there is a significant difference between intelligence and utility. Regarding function scope, utility is intended to evaluate basic functions that are already popular in the current market, whereas intelligence is used to assess new functions that are relatively less widespread and mostly dependent on real-time Internet services. In terms of evaluation procedure, utility evaluates each function separately, whereas intelligence highlights the synergy within a set of consecutive functions in a specified scenario.

The measuring methods are notably different between rational and emotional indexes. The emphasis in rational indexes evaluation is to avoid errors, which implies that there is usually an optimal theoretical value. For example, a 100% success rate in task execution is optimal. Sometimes the optimal value is not actually attainable, but it is indisputable as an optimal upper limit. For example, the minimum impact of interaction tasks on speed and lane keeping is zero. Conversely, the evaluation of emotional indexes emphasize the possibility of improvement. Therefore, there is generally no optimal theoretical value. Full scores in the metric can only be specified manually. For example, theoretically, the higher the number of ecological functions supported by the system the better, and there is no upper limit.

4.3.2 Relatively Subjective Indexes

The concepts rational and emotional indexes, and objective and subjective evaluation are sometimes confounded. Some may assume, in a superficial manner, that rational indexes are objective and emotional indexes are subjective, when in fact there is no necessary correspondence between these two sets of concepts. Rational indexes may also be subjective. For instance, a driver can assign a subjective score to the safety of a driving process. Emotional indexes can also be objective. For example, the icon visibility on a screen can be described objectively in terms of size and color differences relative to the background. Nevertheless, most rational indexes are easier to evaluate using objective criteria, whereas some emotional indexes are not readily assessable using objective criteria. Among the four emotional indexes, most of the elements in cognition and intelligence are easier to evaluate objectively, whereas those in value and aesthetics are generally more difficult to evaluate objectively.

The value and aesthetic indexes are highly subjective. If a group of experts are asked to conduct a subjective evaluation, their opinions will most likely differ. However, the opinion of each expert is still valid, rendering standardization impossible. If a standardized and quantifiable evaluation is needed, a few typical reference cases can be provided as a checklist, against which the product can be evaluated. For example, high-granularity particle elements can reflect an aesthetic tendency toward a sense of technology.

This checklist evaluation approach has two characteristics. First, as the reference case itself is an existing design, it can only guide products toward achieving an acceptable design level in the current market, but not toward pioneering forward-thinking designs. Second, the orientations examined under each index level should be diverse due to the individual differences in subjective indexes; the performances in different aspects are not cumulative. For example, under the aesthetics index, there are different aesthetic orientations, such as a sense of technology and luxury. An HMI system may reach a relatively high level in different aesthetic orientations, or it may attain the highest level in only one certain orientation. However, we cannot say that the former design is better than the latter, only that they have their own characteristics in terms of positioning or that a system performs better in a particular aesthetic orientation.

4.3.3 Inter-constraints Among Indexes

On the basis of such an evaluation index system, is there a theoretically optimal solution where all evaluation indexes are maximized? The answer is probably no, as there are mutual contradictions and constraints among the indexes. Take the Apple iPhone as an example; multiple applications have been designed so that the back button appears in the upper-left corner of the screen, as shown in Fig. 4.13. This design is good from a cognitive point of view. It offers good visibility as it is not easily blocked by the user's hand and its fixed position is easy to recall and locate. However, when the phone is in the right hand, it is difficult to reach the back button with the thumb, which will affect operational efficiency. Automotive HMI systems are more complex than smartphones, which entails that such constraints among indexes are more common. Given the contradictions and constraints among evaluation indexes, designers should consider the positioning of their products, and the target groups in the product design and development stage and select the most appropriate evaluation indexes.

4.4 Differences in Demand Between the Chinese and European Markets

Compared with the European automotive market, the Chinese market had a belated start but grew more rapidly. In 1999, there were only five million privately owned vehicles in China approximately. By 2009, that number had surpassed 50 million, a tenfold increase in only a decade. Moreover, in 2009, China became the world's largest automotive consumer market, and has remained in that position ever since. By 2022, the total vehicle ownership in China had surpassed 400 million. This rapid development has resulted in a lack of deep roots for traditional automotive culture that has dominated Europe and America. Chinese users have a more open and flexible understanding of car culture. People are willing to integrate vehicles with other aspects of their lives and are eager to see automotive technological innovations. Differences in the understanding of vehicles are often magnified when reflected in automotive HMI interfaces.

Many European users view vehicles as serious driving machines, where the driving experience itself is the core of the process. In contrast, Chinese users tend to consider the vehicle not only as a means of transportation but also as a part of their living space, and even as a companion or family member. Automotive HMI can shape such

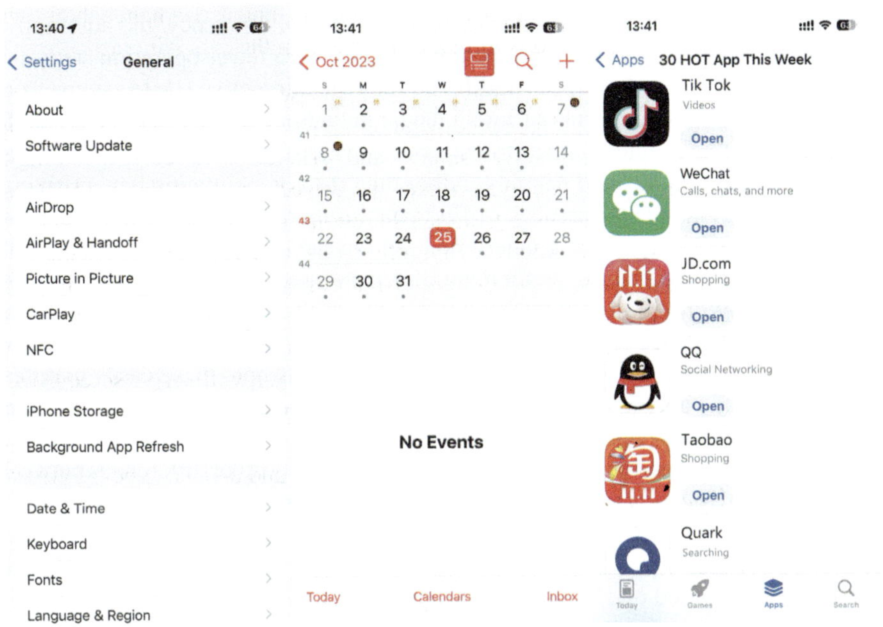

Fig. 4.13 Multiple applications on Apple iPhone have the back button designed to appear in the upper-left corner

relationships, making them more tangible and easily perceptible. For instance, if a voice assistant greets the car owner with a smiling face, it can undoubtedly convey the vehicle's vitality and friendliness more effectively.

With the escalation of electric vehicles in the current market, how do users perceive electric vehicles? Is that perception different from that of internal-combustion-engine vehicles (ICEVs)? If electric vehicles are viewed primarily as a mode of transportation, there is no essential difference between an electric vehicle and an ICEV; they are simply replacements. However, if electric vehicles are seen as intelligent devices, then they clearly set themselves apart from ICEVs. In Europe, many consumers may opt for the former view; whereas in China, they often opt for the latter. Electric vehicles are considered intelligent devices and are a step above ICEVs. Although it is not stipulated that electric vehicles must be intelligent, consumer perception has firmly linked electrification with intelligence. For example, in 2022, some users expect a 200,000 RMB electric vehicle to be able to play videos on the central information display, but they do not have such expectations for an ICEV costing 500,000 RMB or more.

Regarding specific evaluation indexes, there are significant differences in the needs of Chinese and European users as well, and here are a few examples for reference. The unique preferences of Chinese users include the following: First, in terms of utility, the features of the automotive HMI system need to be supported by more diverse interaction modalities to showcase technological capabilities, even if this might exceed the actual user needs. Second, in terms of efficiency, the automotive HMI system should resemble familiar smartphones, with fewer operation steps and the non-requirement to follow the traditional logic of automotive HMI systems from a decade ago. Finally, in terms of intelligence, the automotive HMI system should function more like an intelligent terminal, covering more everyday scenarios, such as movies, games, karaoke, and napping, rather than merely facilitating better driving. In contrast, the unique preferences of European users include the following. First, in terms of safety, they place greater emphasis on the safety risks associated with driver distractions, and hence, prefer to minimize the operational and cognitive loads. Second, in terms of cognition, they favor interaction paths with clear logical structures and sufficient guidance for various interactions. Finally, in terms of aesthetics, they prefer angular lines and vibrant colors. We will discuss how these cases manifest specifically in the automotive HMI system, as well as their pros and cons, in greater detail in subsequent chapters.

In addition to the aforementioned points, there are many other differences between the Chinese and European users in their demands for automotive HMI systems. One of the most significant root causes of these differences lies in the differences in cultural and value orientation. In Chap. 10, we will delve deeper into user value orientations and HMI needs based on Hofstede's six-dimensional (6-D) cultural model.

References

1. Ergonomics of human-system interaction - Part 210: Human-centred design for interactive systems: ISO 9241-210:2019 [S].
2. James R. Lewis, Assessing User Experience (UX) with Two Items: The UMUX-LITE [R], HCII 2019 Tutorial.
3. Tetsuya Tarumoto. User experience and usability testing [M]. Beijing: People's Posts and Telecommunications Press, 2015.
4. Yuval Harari. Sapiens: A brief history of humankind [M]. CITIC Publishing Group, 2014.
5. Benedict Anderson. Imagined Communities: Reflections on the Origin and Spread of Nationalism [M]. Shanghai: Shanghai People's Publishing House, 2005.
6. Fahy J, Jobber D. Foundations of Marketing [M]. McGraw Hill Education, 2015.
7. Road vehicles — Ergonomic aspects of transport information and control systems — Simulated lane change test to assess in-vehicle secondary task demand: ISO 26022:2010 [S].

Chapter 5
Utility

5.1 Development

Utility refers to the capacity of the automotive HMI system to perform specified interaction tasks effectively, accurately, and reliably. It is one of the three fundamental usability metrics, and is derived from the concept of effectiveness outlined in the ISO-9241 definition of usability. Effectiveness refers to the accuracy and completeness with which users accomplish specified tasks [1].

However, the existing scope of effectiveness is inadequate to address all the pertinent aspects related to automotive HMI. Hence, it is crucial to establish a broader scope of usability specific to automotive HMI, which takes into account its three unique characteristics.

First, automotive HMI includes a wide range of interaction tasks, but the availability of these tasks varies across different car models. For instance, certain models allow users to add waypoints to their planned route in the navigation system, while others do not; some models support streaming online music, while others do not. Hence, it is essential to evaluate the availability of specific functions before assessing their effectiveness during use. This disparity in the number of functions is typically not emphasized when evaluating devices other than vehicles. In the case of household appliances such as cooker hoods, the primary functions of various products are generally similar, without noticeable differences. Similarly, smartphones may have a rich variety of features, but disparities in the number of features generally stem from applications (e.g., Google Maps) rather than the operating system itself. Nevertheless, smartphone evaluations generally do not delve into the specifics of individual application software.

Second, a single interaction task can sometimes be accomplished using various interaction modalities, and the modalities offered vary across different car models. For example, some models can only support opening the sunroof with physical buttons, while others offer voice control or the central touchscreen for this task. Thus, the usability of automotive HMI requires examining the richness of matching

J. Ma and Z. Gong, *Automotive Human-Machine Interaction (HMI) Evaluation Method*, Research on Automotive Intelligent Cockpit,
https://doi.org/10.1007/978-981-99-9951-4_5

interaction modalities with interaction tasks. Similar situations are rare for other products. For example, when entering a navigation destination into a smartphone's map application, almost all models can provide both on-screen typing and voice control as interaction modalities with no disparity; hence, there is no need for comparison.

Finally, the usage environment of automotive HMI is highly complex. For instance, the interactive system may be subjected to prolonged exposure to direct sunlight, leading to high temperatures. Additionally, when using voice interaction, the system's microphone may encounter various complex noises that interfere with the user's voice input. Furthermore, in the event of these environmental disturbances, the user is generally unable to remove them proactively; instead, the system has to operate with these interferences. Therefore, the usability of automotive HMI must involve examining the stability of the system in various complex environments. In contrast, the working environment of other appliances and consumer electronic products is comparatively less complex.

5.2 Evaluation Indexes

Usability in automotive HMI evaluation can be divided into second-level indexes, including availability, task success rate, reachability, stability, and modality enhancement.

5.2.1 Availability

Availability refers to the ability of the system to perform an interaction function through a specific interaction modality, such as, the ability to enter a navigation destination via voice control. Some interaction functions do not require direct input from the driver, and their availability does not require considering the interaction modality. For example, when a vehicle is approaching an intersection, lane-guidance information is automatically displayed. Thus, when evaluating the availability of such information display, only the presence of this function needs to be examined, not the interaction modality.

Interaction Tasks and Interaction Modalities

Based on its definition, availability encompasses two levels: the presence of a specific interaction function, i.e., the richness of available functions, and the ability to implement it through a specific interaction modality.

The interaction functions involved in availability cover all basic and extended interaction tasks, as shown in Table 5.1. Currently, most mainstream intelligent vehicles available in the Chinese market generally support over 90% of the tasks listed in the table. Only a few tasks directly related to hardware costs are not available in

Table 5.1 Common basic and extended interaction tasks

Module	Basic interaction tasks	Extended interaction tasks
Media and entertainment	Song search, volume control, next track, switch to a Bluetooth audio source	Play media, add to favorites, play the radio, play videos
Map navigation	Enter a destination and start the navigation, view full route overview, turn off voice guidance, add waypoints, add gas/charging stations along the route, end navigation	Frequently used destinations, check points of interest (POI), switch to detailed guidance, access traffic information, switch to 2D map
Telephone	Enter a number and start a call, dial a designated contact's number, answer a call	
Climate control	Turn off the climate control system, temperature control, fan speed control, switch to external circulation	
Vehicle control		Turn on seat heating, turn on defogging, adjust fan direction, unlock the doors, open the trunk, turn on high beam headlights, switch between driving modes, open the left front window, open the right rear window, open the sunroof, open 50% of the sunroof, shut the sunroof, turn off the central information display, switch the ambient light color

low-end products, such as seat heating and ambient lighting. Therefore, the richness of interaction tasks varies only slightly among different products.

For the same interaction task, the interaction modalities available in different car models can differ significantly. Some models are more inclined to offer a wide variety of interaction modalities for each task, which provides the user with different choices; whereas other models tend to focus the interaction modalities on the central touchscreen and voice control, minimizing the number of physical buttons. The selection of design strategies is related to their interior styling concepts.

Most of the interaction tasks listed in Table 5.1 are output by the HMI system itself, with the information presented on-screen or conveyed verbally. The absence of these functions represents a deficiency in the HMI system. For example, if the navigation system does not support adding waypoints, the task score for waypoint addition will be affected, regardless of whether the touchscreen or voice control is used. However, the output of certain tasks involves automotive appliances beyond the scope of HMI system evaluation, and hence, determining the presence or absence of these appliances also lies beyond its scope of evaluation. For example, if a vehicle has a sunroof that cannot be opened using voice control, the test item of opening the sunroof using voice control will be marked down. Conversely, if a vehicle does not have a sunroof, the test item for opening the sunroof using voice control will be removed, and no further score deductions will be discussed.

Information Display

Aside from basic and extended interaction tasks, the richness of functions also involves some information display, including the contents of navigation information display and the position of driving information display.

There is significant variation in the richness of information display related to map navigation among different vehicle models. This disparity can be attributed to the complex nature of map navigation, which demands a high level of information richness and timeliness. Additionally, such information relies on robust in-vehicle sensors and cloud data for effective support. Since 2010, the rapid development of navigation applications on smartphones have caused many users to prefer smartphone navigation over in-vehicle navigation. To gain user preference, automotive HMI systems need to transform the navigation experience while avoiding mere duplication of smartphone navigation functions. With the central information display of mainstream intelligent vehicles surpassing 10 inches in size, providing at least three times the screen area of a smartphone, there is ample space for innovation within in-vehicle navigation.

The interface of in-vehicle map navigation systems can provide driving guidance to users in addition to basic information such as road location, driving guidance, and distance to destination. Some examples are as follows:

(a) Driving guidance for consecutive intersections. As city roads become increasingly complex, situations where two intersections are in close proximity often arise, as illustrated in Fig. 5.1. After leaving the first intersection, if the driver knows which lane to enter only at the second intersection, there may not be sufficient time to react. Even a delay of several seconds in the navigation system might cause the vehicle to enter the second intersection without receiving proper guidance, significantly increasing the likelihood of taking the wrong route. Therefore, displaying guidance information for the next two or more intersections can better prepare drivers, as shown in Fig. 5.2. Perhaps due to the limited screen size, mainstream smartphone navigation applications in China were unable to provide guidance beyond the second intersection before 2020. Therefore, some smartphone projection navigation applications and in-vehicle navigation systems based on smartphone applications are not equipped with such information.

(b) Dynamic road rules. There are four common situations where road rules may change. The first is dedicated lanes during fixed time periods, such as bus lanes during the morning and evening rush hours. The second involves tidal lanes, where the travel direction changes at different times, e.g., from west–east to east–west, as shown in Fig. 5.3. The third is the variable lanes at intersections; for example, the second lane from the left can sometimes be used for turning left and sometimes for going straight. Finally, the fourth refers to lower speed limits on highways in slippery or poor visibility conditions. The rules in the

Fig. 5.1 Fork intersection approximately 25 m away from the ramp entrance of the Inner Ring Elevated Expressway and the North–South Elevated Expressway interchange in Shanghai, China

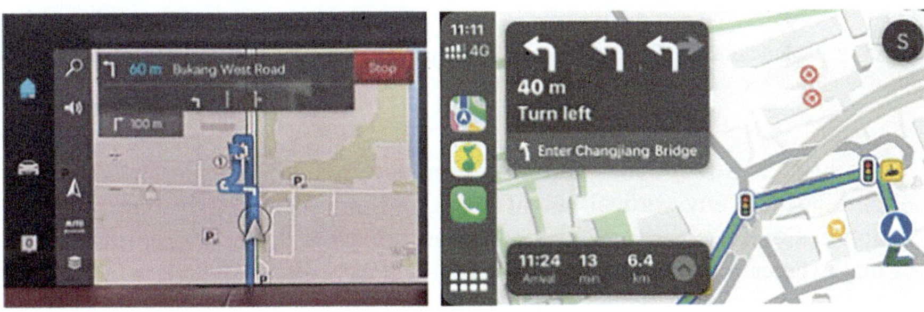

Fig. 5.2 Porsche Taycan and Apple Maps (CarPlay projection) providing guidance for the second intersection (captured in 2021, translated from Chinese language)

first case are fixed and easier to display dynamically in the navigation system. However, there are no visible fixed rules for the other three cases; therefore, real-time collaboration between navigation systems and road traffic authorities, or advanced big data prediction are needed in order to provide real-time dynamic information.

(c) Vehicle route deviation. When the vehicle deviates from the planned route, the navigation system needs to recalculate the route promptly; otherwise, the originally planned guidance content will cause erroneous interference to the driver. Route deviations in the horizontal direction can be easily detected through satellite positioning, whereas deviations in the vertical direction are difficult to detect using satellite positioning alone. For example, if the vehicle mistakenly enters the off-ramp of an elevated expressway too early, then high-accuracy acceleration sensors are required to identify whether the vehicle is accelerating vertically and determine if it is going uphill or downhill. Some vehicles with superior performance can detect that they are going downhill when the vertical

Fig. 5.3 Tidal lanes on the Garden Bridge in Shanghai with real-time guidance of driving direction using dynamic arrows

height decreases by 3 m, while the navigation systems of other models are unable to determine any vertical height changes. Smartphone navigation projection applications, such as CarPlay, use low-accuracy internal accelerometers, which makes determining the vertical vehicle height difficult, and results in the inability to actively identify related route deviations.

In the future, with further technological developments, navigation systems should aim to provide more accurate and immersive information.

(d) Augmented reality (AR) navigation. AR is a technology that seamlessly blends virtual information with the real world. In the context of navigation, AR can superimpose guiding signs directly onto the real road, enabling drivers to intuitively determine the correct route, without the need for cognitive interpretation of abstract navigation icons. There are two primary methods for presenting AR navigation. The first approach involves displaying the AR information on either the instrument cluster display or central information display. The camera captures the road ahead and displays it in real time on-screen, while superimposing the guiding signs, as shown in Fig. 5.4. This type of AR navigation has a large field of view (FOV) and the guiding signs exhibit a good visual sense of alignment with the ground. However, the driver has to lower their head to see the navigation interface. The second method involves projecting the guiding signs onto the HUD within the driver's forward view, aligning them with the real road, as shown in Fig. 5.5. With this approach, drivers can maintain their focus straight ahead, minimizing visual distractions. However, currently, the FOV of the HUD screen can only reach approximately 12°, and establishing a visual sense of alignment with the ground using the guiding signs is difficult.

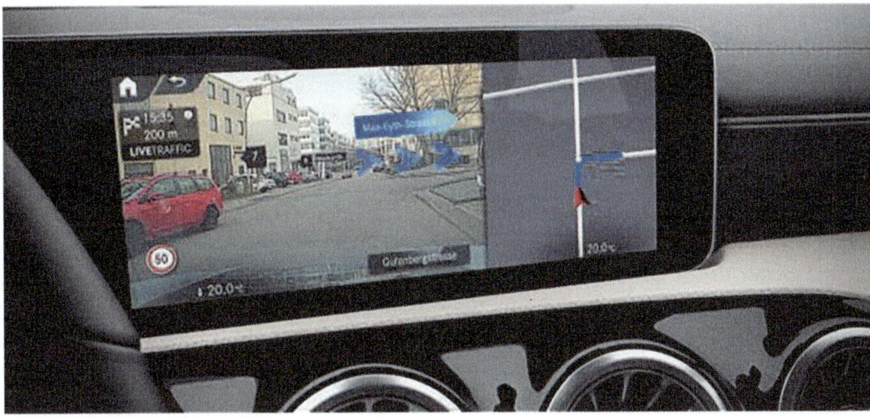

Fig. 5.4 AR navigation on the central information display of the Mercedes-Benz A-Class (2019) (*Source* The Mercedes-Benz Group)

Fig. 5.5 AR navigation on the HUD of the Mercedes-Benz S-Class (2020) (*Source* The Mercedes-Benz Group)

(e) Real-time lane information. Current mainstream navigation systems are only capable of providing driving lane suggestions instead of lane change recommendations based on the lane in which the vehicle is currently traveling. This is because they cannot accurately determine in which lane the vehicle is at any given moment. In future, high-precision positioning and visual recognition technologies can be used to accurately identify the location of the vehicle and provide lane change suggestions, which will not only improve the navigation system's performance but also expand the boundaries of autonomous driving.

Notably, availability focuses solely on the richness of information and does not consider the specific methods of presentation. For instance, a detailed and magnified view of a real intersection may not necessarily provide more essential information compared to a simpler intersection image. In such cases, the former primarily enhances the cognitive comprehension of elements rather than improving utility.

Additionally, ecological service functions are not categorized under availability but rather fall under the intelligence index. This is because ecological service functions have a wide range of applications, and undergo rapid iterations and updates. Furthermore, the necessity of certain new functions is still inconclusive. In other words, it is better to have more items that can be implemented within the availability category, but the necessity of certain new ecosystem functions, such as hotel room reservations within the HMI system, may be debatable. Therefore, we cannot state definitively that the absence of such a function signifies a lack of utility.

5.2.2 Task Success Rate

The task success rate (TSR) refers to the ratio of the number of successfully and accurately completed interaction tasks to the total number of operations in the automotive HMI system. There are many reasons for task failure or errors, which may occur during user command input or during the command recognition and execution by the HMI system, as shown in Fig. 5.6.

Incorrect commands by the user are manifested as incorrect interaction paths, where the user does not complete the interaction task in the correct step sequence but instead taps the wrong icon, presses the wrong button, or says the wrong voice command. Sometimes, the user can go back to the previous step and continue with the correct operation after detecting an operation error, while other times they have to start over. The reason for an incorrect interaction path is usually because the user cannot clearly see or comprehend the icons or texts on the interface, and does not know how to perform the correct operation, which implies that they can only resort to trial and error. For tasks frequently performed while driving, it is necessary to ensure that most users should be able to achieve 100% accuracy for the interaction paths. Conversely, interaction tasks that are less frequently used and highly complex

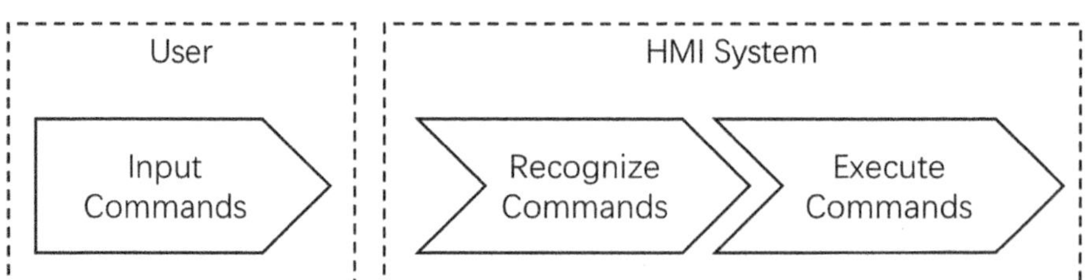

Fig. 5.6 Possible stages of error occurrence during task completion

are typically hidden deep within the directory hierarchy, making them challenging to locate and remember. Therefore, it may be reasonable to relax the requirement for accuracy in their interaction paths, such as when switching the measurement units of driving information.

The HMI system should successfully and correctly recognize different user command inputs. For physical buttons, it is not difficult to correctly recognize each button press by the user, unless there is a mechanical failure with the button itself. For touchscreens, input recognition failures may occur due to factors such as the user's finger deviating from the intended position, tap duration being too short, or finger sliding during the tap. To reduce the likelihood of these failures, it is necessary to design appropriate icon sizes, as well as increase the screen hardware sensitivity and fault tolerance. For voice control, there may be instances of system activation failure if the task involves voice activation with specific words or speaking directly without a activating word, whereas using a dedicated voice control button or icon to activate the system usually does not result in failure. After the user utters specific commands, the speech recognition system may fail to recognize each word spoken by the user due to unclear sound collection, non-standard pronunciation, unstable network connection (speech recognition generally requires cloud support), insufficient semantic analysis capability, or incomplete vocabulary.

Once the HMI system recognizes the user's command, it must execute the command correctly. For buttons and touchscreens, correctly executing recognized commands is not difficult unless the system experiences severe delays or crashes. For voice control, failure to complete the task or incorrect task execution may still occur even if the system correctly recognizes every word spoken by the user. For instance, the vehicle's HMI system may understand the user's command to "open the sunroof", but it may not support the execution of this task through voice control. In such cases, the system can only respond with a message such as "voice control currently does not support opening the sunroof". This is because the input range for voice control is open and unrestricted, while buttons and touchscreens have defined boundaries, limiting the range of tasks that the vehicle system can perform.

To differentiate between a system's failure in command recognition and task execution, the system needs to display the command recognition status. For example, after tapping a touchscreen icon, the icon changes color or makes a ticking sound, or after the user speaks the voice command, the recognized text is displayed on the screen. However, such recognition status displays are not available in some car models; therefore, we can only integrate the recognition and execution success rates for the evaluation sometimes.

Furthermore, there are also success rates for tasks that do not involve user input, such as the ability of the navigation system to identify route deviations via satellite positioning and accelerometer sensors.

5.2.3 Reachability

Reachability refers to the ease with which the position of the buttons or screen areas that need to be reached by the users' fingers, when operating specific tasks. If these positions are too far from the users' body, they have to lift their shoulders forward or move their bodies to the right for a successful operation. This not only affects the convenience and comfort of the operation, but may also cause instabilities in vehicle steering wheel angle, thus creating safety hazards. During actual evaluations, reachability mainly targets touchscreen interaction because the icon reachability on the central information display in many car models is inadequate and thus requires greater attention. In contrast, the central console buttons in most car models are naturally located closer to the driver, where reachability usually does not pose a problem.

The J287 standard of the Society of Automotive Engineers International (SAE International) describes a recommended measurement method for the driver's hand-control reach, as shown in Fig. 5.7 [2]. The three-dimensional orthogonal coordinate system established in the figure considers the H-point (center of the hip joint) as the origin (for cars with a high seating position or tall drivers, the origin will be adjusted forward from the H-point) [3]. Each node on the grid in front of the driver has fixed Y- and Z-coordinate values, and the X-coordinate value (positive towards the front of the car) of each node can be obtained by consulting a table. The inboard and outboard curved surfaces composed of all nodes correspond approximately to spherical surfaces with the shoulder joint as the center. The reference value of the X-coordinate depends on three factors. The first is the driver's hand posture during operation. The benchmark value is obtained by the driver gripping a 25-mm-diameter knob with the thumb, index finger, and middle finger. If the driver taps with one finger, the X-coordinate reference value can be extended by 50 mm; and if the driver holds it with the palm, it is reduced by 50 mm. The second factor is the vehicle seating position. The seatback of a vehicle with higher seating position (e.g., an SUV) is usually more upright than that of a vehicle with a lower seating position (e.g., a sports car). Therefore, it has a larger X-coordinate reference value. The third factor is the height of the driver. The reference value in the SAE J287 standard is within the reachability of 95% of the American driving population, where the male-to-female ratio can be adjusted according to the specific vehicle position, and is generally 50:50, 75:25, or 90:10. The higher the average height of the group, the larger is the X-coordinate reference value.

There are four points to consider when adopting this measurement method. First, reachability by 95% of the driving population does not refer to the area reachable by taller individuals in the 95th percentile for height, but rather the area also reachable by shorter individuals in the 5th percentile. Misinterpreting this can lead to poor reachability design of the central information display. Second, the reachable area refers to the reachable range when the driver leans forward with the seat belt on, which can affect stability during driving. Therefore, areas that are far away from the driver within this range will still affect driving safety. Third, the distance between

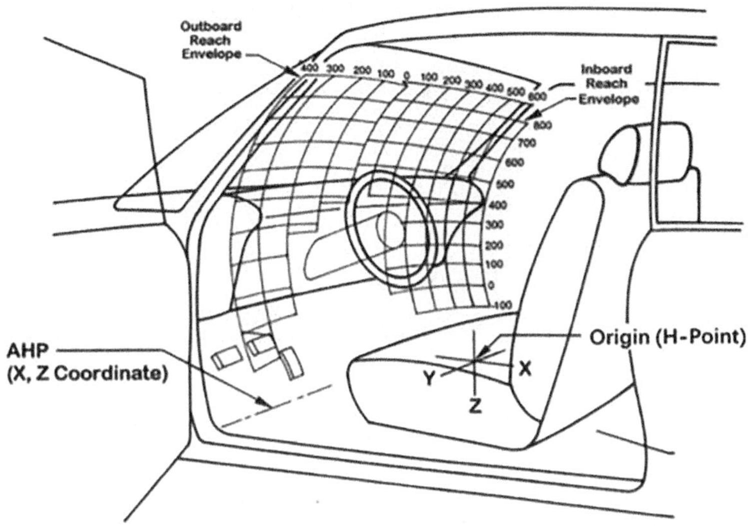

Fig. 5.7 Recommended measurement method for the driver's hand-control reachable area (*Source* SAE J287 Driver Hand-Control Reach)

the H-point and accelerator heel point (AHP) varies for people with different heights [4]. Finally, the SAE J287 standard is defined based on the physique of American drivers. When testing for the other market, adjustments should be made based on the physique of these drivers.

The SAE J287 standard plays a vital role in the automotive design modeling process. However, it is relatively complex to reproduce such three-dimensional coordinates in the cabin when testing a real mass-produced vehicle. A simpler testing method is the spherical coordinate measurement method. Using the driver's shoulder joint as the origin, arcs of different lengths are drawn on the central information display, which represents the intersection between the spherical surface of the hand reachable area and the plane on which the screen is located. By observing the positional relationship between the icons on the screen and arcs drawn, the range of distances between the icons and the driver's shoulder joint can be determined. Based on the equivalent conversion of the SAE J287 standard, when using the spherical coordinate measurement method, the maximum radius reachable by the driver is approximately 900 mm, with the shoulder joint resting on the seatback as the center of the sphere. The reason this distance exceeds the driver's arm length is because when using the touchscreen, the driver's shoulder will leave the seatback as they lean forward, or even move their entire upper body forward and to the right. To ensure both comfort and driving safety, a distance range of 800–850 mm would be more suitable. Taking the Mercedes-Benz EQC as an example (Fig. 5.8), when performing the task of entering a destination, all icons are located within a range of 840–850 mm from the driver. However, if an icon appears in the upper-right corner, its distance from the driver will exceed 880 mm.

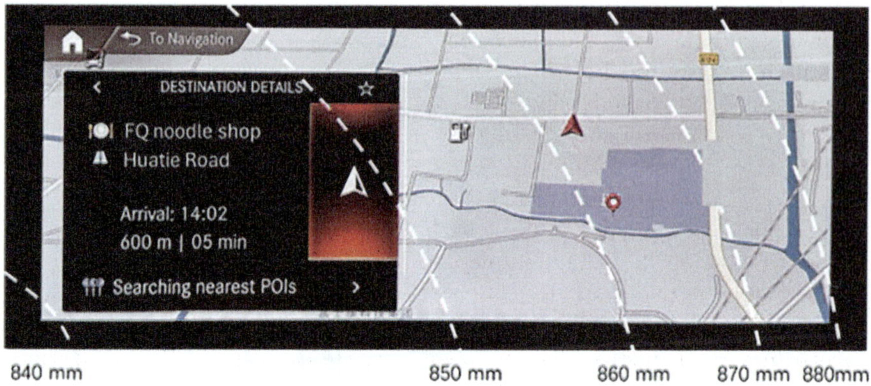

Fig. 5.8 Schematic of central information display reachability in the Mercedes-Benz EQC (2019)

Regardless of the measurement method, the position of the H-point is crucial. Errors in the anterior or posterior direction of the H-point will be fully reflected in the measurement deviations. Therefore, the H-point requires meticulous measurement and positioning.

5.2.4 Stability

Owing to the highly complex working environment of automotive HMI systems, a system that runs smoothly in the laboratory may not necessarily be able to handle all real-world environments in a stable manner. Automotive HMI systems not only need to deal with complex lighting conditions and noise, but also enhance their own heat dissipation performance and stability, as reflected in the following aspects:

(a) Anti-reflective performance. Reflections on the screen can affect the user's ability to read the content displayed, which will prevent them from completing the interaction tasks correctly, safely, and efficiently. Excessive sunlight reflecting on the screen can even hinder the driver's view of the road ahead. Sunlight can enter the vehicle from all directions, including the rear window, side window, or sunroof. Additionally, reflections on the screen may occur due to the passengers' faces or light-colored interior elements. Many car models before 2010 installed transparent downward-sloping concave covers around the central information display and instrument cluster display to filter the reflections of external light, as depicted in Fig. 5.9. However, with larger screen sizes and the addition of touch functionality, such designs are no longer practical. Currently, screen anti-reflective performance can be enhanced through surface coating treatment, adjusting the pitch and tilt angle, and using a concave curved screen, as shown in Fig. 5.10.

Fig. 5.9 Downward-sloping transparent cover outside the central information display and instrument cluster display of the BMW 5-Series (2004) (*Source* The BMW Group)

Fig. 5.10 Concave curved instrument cluster display of the Porsche Taycan (2020) (*Source* Porsche AG)

(b) Anti-noise performance. Voice control should work properly even in the presence of noise in the cabin. Anti-noise performance is mainly influenced by three factors. First, cabin noise is extremely complex, comprising of engine/motor, tire, and airflow noises. It is unlikely that users will slow down deliberately to improve the speech recognition rate, which makes it challenging to attenuate these noise sources. Second, microphones in the cabin are typically positioned at the front of the roof or in the dashboard, which is at a certain distance away from the user's mouth. Often unaware of the microphone's location, users will not actively move closer and speak into the microphone, as one would do when using a telephone. Additionally, sound from the audio system and speech from other passengers can also interfere with speech recognition.

(c) Heat dissipation performance. Under direct summer sunlight, the dashboard can be heated to a temperature exceeding 70 °C, while the HMI system hardware under the dashboard will also continue to generate heat. The superposition of the two can further increase the operating temperature of the system. Excessively high temperatures can affect the operational stability and service life of the system hardware. Additionally, elevated temperatures of touchscreen surface can make the user feel uncomfortable when tapping it. Therefore, good heat dissipation performance is extremely important for automotive HMI systems.

(d) Minimal system crashes. The HMI system should run without experiencing system crashes or automatic reboots during use. Such issues can cause user anxiety and frustration, leaving a highly negative impression of the overall product performance. If a crash or reboot affects the display of navigation guidance information, it can mislead drivers into taking wrong routes. Similarly, if it impacts the display of driving information, driver assistance features, and so on, it may lead to traffic violations or accidents.

5.2.5 Modality Enhancement

With the rapid development of automotive HMI, various interaction modalities may intersect and influence each other. These cases cannot be easily categorized into the four typical interaction modalities defined in Sect. 3.2, nor are they sufficiently substantial to form a new interaction modality. Therefore, we propose a second-level index, modality enhancement, to group these related issues together. The specific content encapsulated by modality enhancement will continue to expand as technology progresses. For now, the discussion will focus primarily on the following two points.

The voice interaction modality should be both independent and complementary. Unlike touchscreens or buttons, voice control is the only modality that does not require any physical movement from the user. When using voice to perform operations, users naturally expect to complete all steps of a task using voice alone. However, some vehicle models do not follow this approach in their design. For instance, after inputting the navigation destination and selecting a driving route using voice control, some models require users to manually tap the "Start Navigation" icon on the screen instead of using voice to initiate navigation. This lack of independence in voice interaction disrupts the user's coherent understanding of interaction modalities and causes inconvenience. While every step of a voice interaction task can be controlled by voice, this does not mean that every step must be voice-only. In the voice interaction process, users should be allowed to use the touchscreen for some steps, so that the advantages of voice control and touchscreens are mutually complementary. When the HMI system presents a list of choices based on the user's voice input, it may be less efficient for users to verbally state their selection or the corresponding number. In this case, some users might prefer to use the touchscreen to directly tap their selection.

The novel "see and talk" interaction method was pioneered by Xpeng Motors in 2021, and has been emulated by other manufacturers. Using this method, users can carry out interactions through voice control by directly reading out the content displayed on the screen without tapping. For instance, when a Li Auto One user says "Top Chart" on the music interface, a small dynamic blue circle will appear on the corresponding icon, which then leads to the song list, as shown in Fig. 5.11. Although "see and talk" involves only voice input without any tapping, the interaction path it follows is shown entirely on the central information display, which requires step-by-step rather than direct voice control to access deeper tasks. Therefore, "see and talk" is a form of extended voice control for touchscreen interaction enhancement. It allows the driver to operate the central information display without taking their hands off the steering wheel, which avoids unstable steering wheel grip when reaching out to touch the screen, thereby improving safety. This is especially true for the icons that are farther away from the driver. However, it takes longer for the user to read out the content on the screen than it does to tap the icons with a finger, and it also takes longer for the system to recognize and act on the voice command than it does to respond to touchscreen input. Moreover, "see and talk" does not reduce the interaction steps as typical voice control does. Therefore, due to its inherent disadvantages in efficiency, "see and talk" can only serve as a supplementary interaction modality in the short term and is unlikely to become the dominant mode of interaction in the long term.

Fig. 5.11 "Top Chart" selection in the music interface using "see and talk" interaction modality in the Li Auto One (2021), translated from Chinese language

5.3 Summary of Evaluation Results and Design Suggestions

5.3.1 Product Positioning and Design Strategies

When comparing vehicles, many performance indexes show a positive correlation with vehicle price, leading to the division of vehicles into different market segments. Typically, higher-priced products offer superior power performance, handling, and comfort. Nevertheless, this rule is not prominent for automotive HMI. In the current Chinese market, all vehicles priced above 20,000 Euros and even some vehicles priced above 13,000 Euros that prioritize intelligence are directly comparable in terms of HMI utility. This is because the intelligent cockpit is an important selling point for almost all cars, and no major manufacturer is willing to lag behind its competitors in this area. This is particularly true for relatively inexpensive local Chinese automotive brands, whose intelligent cockpit-related software and hardware account for a higher percentage of the vehicle price than traditional international brands. Additionally, when the basic hardware for automotive HMI reaches a certain level, further upgrades mainly rely on software, and once the software is developed and perfected, it has a very low marginal cost. In other words, models with a high sales volume and low price may have comparable or even greater investment in software development than those with a low sales volume and high price. In addition to utility, other first-level evaluation indexes have a weaker correlation with cost; therefore, there is generally no need to divide them into different price segments for comparison.

The selection of interaction modality with respect to availability is closely related to the model design strategy. For example, Tesla Model 3, which was released in 2017, did not have any central console buttons, nor did it have a traditional instrument cluster display. Instead, its functions were incorporated into a central information display, as shown in Fig. 5.12. It is an ongoing trend to continuously simplify the physical hardware of HMI. As a "visible" interaction modality, larger screens and more buttons contribute to higher manufacturing costs; however, cost-saving is not the sole reason for this trend. Physical buttons occupy the space of other elements owing to their shapes; in particular, they clash with large vertical screens in terms of layout positioning. On the other hand, physical buttons have fixed functions, which cannot be updated through later software upgrades. Therefore, although having fewer interaction modalities may result in lower availability index scores, it does not necessarily mean that such a design is incorrect. Its performance in terms of usability, including safety, efficiency, and cognition, should also be considered.

For the "invisible" interaction modality of "voice control," it is better to have a broader scope of availability. In a market where voice control has been rapidly embraced by users, rich voice interactions have become a mandatory feature for all car models, eliminating the need to discuss whether voice control enhances user value. Feature-rich voice interaction adds little to the hardware cost and does not occupy space that could be used for other in-vehicle elements. Therefore, voice control that is lacking in functions is generally not the result of a deliberate interaction design

Fig. 5.12 Tesla Model 3 (2017) without central console buttons and instrument cluster display (*Source* Tesla, Inc.)

strategy but rather one of insufficient research and development investment by the manufacturer.

5.3.2 Touchscreen Limitations

Owing to their unique advantages, touchable central information displays have become the core interaction modality of current automotive HMI. Screens are capable of displaying infinite content within a finite area through information folding. As the content on the screen can be accessed directly by tapping, users can experience an intuitive "what you see is what you tap" effect. In addition, large screens themselves embody a sense of technology in the cabin, which is also an important factor considered by consumers when buying a car.

However, there are still several challenges in the design of the central information displays with respect to utility. First, the success rate of tapping the central touchscreen is lower than that of pressing traditional physical buttons, which means a lower recognition rate of the user's command input. The first reason for the low recognition rate is the lack of precision in where the user's finger taps. For a majority of vehicles, when users tap on the central information display, there is no support point for the elbow and wrist. Users need to rely on their shoulders as the pivot point to control the movement of the entire arm, which means a slight angular deviation may cause the fingertip to deviate from the intended contact point. It is particularly difficult to control the arm on bumpy roads. Users do not have this problem when using smartphones because the phone is fixed relative to the palm, and the user simply needs to control the thumb with its base as the pivot point and a force arm of only a few centimeters long, which is conducive to tap accuracy. It is not a significant issue with physical buttons either, as the edges of most buttons have distinct edges or contours. Users can judge whether they have touched the center of the button with their fingers and then make adjustments accordingly before pressing it. The second reason is the misjudgment of on-screen gestures, especially on bumpy roads. When the user taps the screen, the finger may slide a small distance on the screen, which is misjudged by the system as a swipe gesture. Whereas, when a user swipes continuously on the screen, their finger might lift off the screen briefly, which the system misidentifies as multiple swipes or taps. Therefore, although swiping can sometimes enhance the interaction efficiency, it is also preferable to operate a task that can be achieved by swiping on the central information display through tapping. Otherwise, it may affect the success rate of tasks such as adjusting the air-conditioning temperature. The third reason is the imprecise swiping operation. The absence of vibrational feedback from the screen prevents users from accurately gauging the distance their finger has moved based on touch alone. For instance, if a 0.5 cm scroll on the screen corresponds to a 1 °C increase in climate temperature, achieving a precise 3 °C adjustment would require users to accurately control a 1.5 cm scroll, which can be challenging. One possible solution is to reduce the scrolling rate of the finger. Let us take the Xpeng P5 as an example, where the user needs to scroll 2.5 cm to increase the climate temperature by 0.5 °C, as shown in Fig. 5.13. This design restricts users from making drastic changes to the temperature with a single swipe, which is typically not a necessary operation for users.

Another challenge for the central information display is reachability. Many models have poor performance in this regard for three main reasons. First, the central information display of some models is not designed to be touchable. For example, the central information display of the sixth-generation BMW 3 Series, which was launched in 2012, can only be controlled by the iDrive knob instead of a touchscreen. Consequently, the display is positioned closer to the windshield without considering reachability, as depicted in the top panel of Fig. 5.14. Although the BMW 3 Series was later upgraded to incorporate a touchable central information display, it was difficult to make dramatic adjustments to the interior layout of the vehicle, which entailed that the central information display was still positioned far from the driver. By contrast, the central information display of the new-generation BMW 3 Series, which was launched in 2019, was initially designed to be touchable and therefore positioned closer to the driver, as shown in the bottom panel of Fig. 5.14. Second, as the central information display becomes increasingly larger, the right side of the screen is bound to be farther away from the driver. If the large horizontal central screen is not tilted toward the driver, reaching icons on the right side becomes challenging. Third, reachability is not the only factor that should be considered in screen design, as the aesthetic of the interior layout is also important. In some car models, the central information and instrument cluster displays were placed on the same plane, aiming to achieve a styling design with a sense of wholeness. However, this design also results in the central information display being positioned further away from the driver, as depicted in Fig. 5.15.

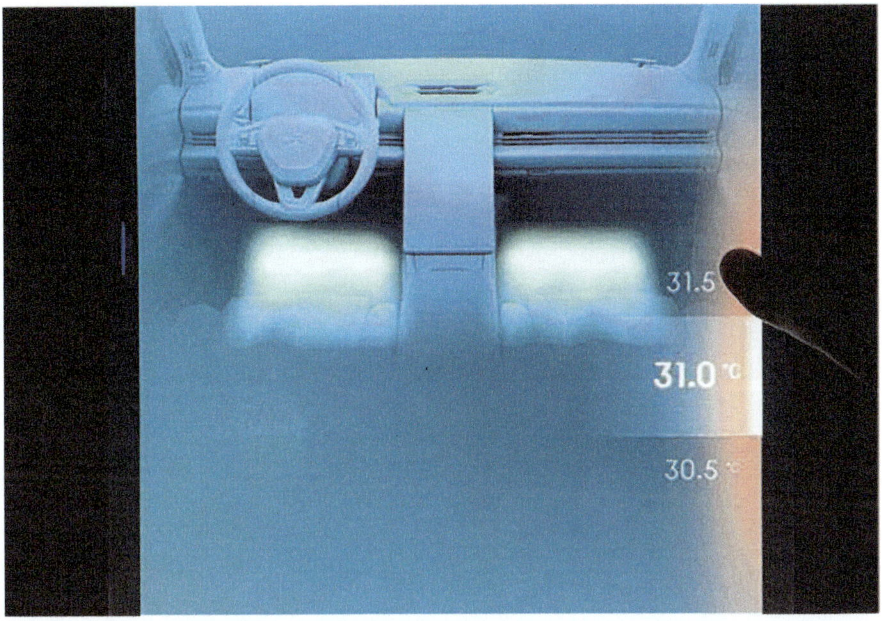

Fig. 5.13 Climate control interface of the Xpeng P5 (2021), in which the temperature changes by 0.5 °C when the finger scrolls approximately 2.5 cm

5.3.3 Button Development

As a well-established modality in automotive HMI, physical buttons have undergone changes in response to digitalization.

Some physical buttons resemble screens in appearance and tactile feedback. The difference between traditional physical buttons and on-screen icons is that each button has an independent tactile boundary (typically, a tiny gap) and an independent downstroke, whereas icons have neither. Buttons that have independent downstrokes but not tactile boundaries are referred to as pressure-sensitive buttons, as depicted in Fig. 5.16. These pressure-sensitive button modules are widely used in household appliances such as rice cookers. They appear relatively simple and provide a pressing feel similar to conventional buttons. Buttons that have neither independent downstrokes nor independent tactile boundaries are called touch-sensitive buttons, as shown in Fig. 5.17. Similar to the operating feel of buttons on induction cookers, basic touch-sensitive buttons do not have any tactile feedback. However, with a vibration motor installed beneath the touch panel, they can simulate the pressing feel of traditional buttons, such as the Home button on the iPhone 6. Despite their concise appearance and better visual sense of technology, both pressure-sensitive and touch-sensitive buttons have abandoned the natural advantages of traditional buttons in task success rate. Although more advanced technologies can address these limitations,

Fig. 5.14 Central information display of the BMW 3 Series (2019, bottom panel), which is positioned closer to the driver than in its predecessor (top panel) (*Source* The BMW Group)

Fig. 5.15 Central information, instrument cluster, and front passenger display of the Li Auto One (2020), positioned on the same plane (*Source* Li Auto, Inc.)

Fig. 5.16 Pressure-sensitive buttons in the Cadillac CT6 (2017) (*Source* General Motors Company)

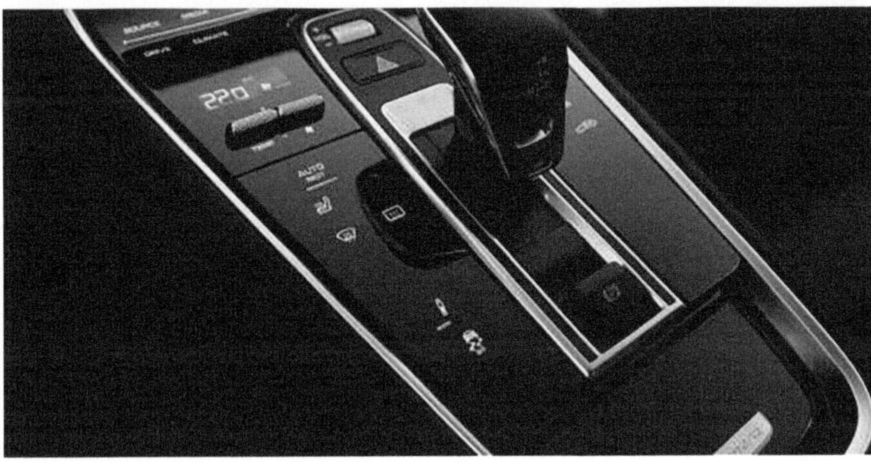

Fig. 5.17 Touch-sensitive buttons with vibration feedback in the Porsche Panamera (2017) (*Source* Porsche AG)

they do not surpass traditional buttons in terms of usability. Therefore, when selecting between push and touch buttons, there is a trade-off between style and utility.

Some physical buttons have been integrated with the screen to overcome the limitations of conventional single-function buttons. For example, the Ford Mustang Mach-E, with a physical knob on the central information display, is capable of performing different functions based on different contents displayed on the screen, as shown in Fig. 5.18. This design not only maintains the feedback feel of traditional buttons but also allows screen-like function folding. However, such designs have not gained widespread popularity in automotive HMI. It is costly to manufacture such a specialized knob or button, which may not necessarily lead to a favorable perception of users. Conversely, knob integration on the screen limits the functional expansion of the screen area to a certain extent. As in the case of the Ford Mustang Mach-E, it is

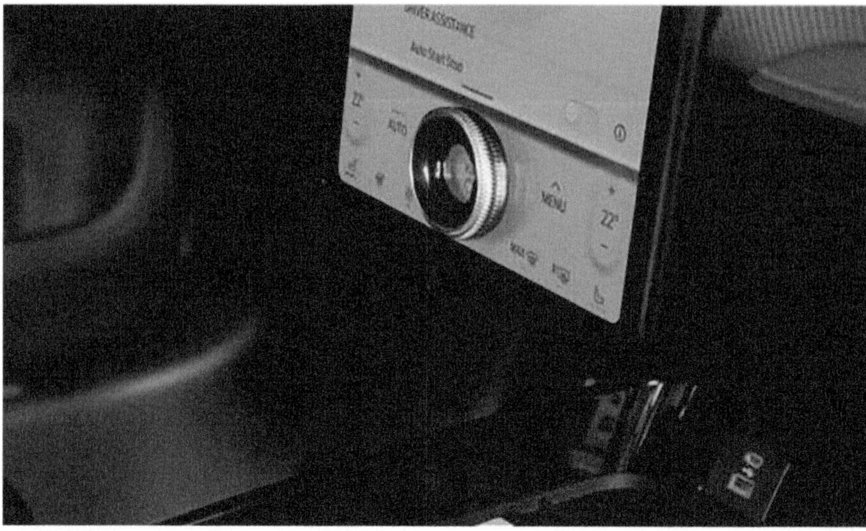

Fig. 5.18 Physical knob on the central information display of the Ford Mustang Mach-E (2021) (*Source* Ford Motor Company)

evident that extending the map page to the inside and both sides of the knob would not be suitable.

References

1. Ergonomics of human-system interaction - Part 210: Human-centred design for interactive systems: ISO 9241-210:2019 [S].
2. Driver Hand Control Reach: SAE J 287-2007 [S].
3. Devices for Use in Defining and Measuring Vehicle Seating Accommodation: SAE J826-2008 [S].
4. Motor Vehicle Driver and Passenger Head Position: SAE J1052-1997 [S].

Chapter 6
Safety

6.1 Development

Safety refers to the ability of the automotive HMI system to suppress driver distraction and enhance driving safety while the driver is performing automotive HMI tasks during driving. Unlike other HMI evaluation indexes, safety is an evaluation index that is unique to automotive HMI. This is because most frequently used tasks in automotive HMI are typically secondary tasks that users need to perform while driving without significantly impacting driving safety. In contrast, for devices such as cell phones or computers, users can usually operate them with full attention without secondary tasks; therefore, there is no need for safety evaluation. Safety plays a special role in automotive HMI system evaluation as it does not evaluate the system itself, but rather the impact of it as a set of secondary tasks on another set of primary driving tasks. The balance between secondary and primary tasks is a major challenge in the design and evaluation of automotive HMI.

6.1.1 Driving Safety and Secondary Tasks

Safety is one of the most important issues in road traffic, and driver distraction has been identified as a significant factor leading to traffic accidents, including vehicle collision [1, 2]. Various studies conducted by researchers worldwide have revealed that distracted driving accounts for 8.3% to 29% of road traffic accidents [3, 4]. Notably, the study by McEvoy demonstrated that secondary tasks were the primary source of driver distraction (68.7%), nearly as prevalent as inattention (71.8%), and significantly more common than viewing outside people, objects, or events (57.8%), talking to passengers (39.8%), drinking (11.3%), eating (6.0%), or smoking (10.6%) [5]. Furthermore, Huemer's comparative studies across multiple countries from 1999 to 2015 revealed a consistent upward trend in the influence of secondary tasks on

J. Ma and Z. Gong, *Automotive Human-Machine Interaction (HMI) Evaluation Method*, Research on Automotive Intelligent Cockpit,
https://doi.org/10.1007/978-981-99-9951-4_6

driver distraction [6]. In 2013, Metz conducted a naturalistic driving study (NDS) in Germany, analyzing 370,000 km of CAN bus data and 20,000 km of video data. The study found that drivers dedicated approximately 40% of their driving time to secondary tasks [7]. However, according to Sayer's analysis of 1,440 in-vehicle video recordings in the United States in 2005, the time spent on secondary tasks accounted for approximately 34% [8]. Although similar large-scale NDSs are not common in China, it can be inferred from the enthusiasm of Chinese people for various consumer electronics that Chinese car drivers currently spend significantly more time on secondary tasks than European and American users did several years ago.

Secondary driving tasks include using hands-free phones, sending text messages, using steering wheel or central console buttons, inputting information on the central information display, visiting websites, and playing games [9]. In 2013, it was reported that half of the secondary tasks related to information input performed by German drivers involved using the automotive HMI system, while the other half were performed on cell phones [7]. With the growing capabilities of automotive HMI systems in recent years, an increasing number of drivers have opted to use the HMI system rather than cell phones to perform secondary tasks. Taking the in-vehicle navigation system as an example, in 2018, only 19% of Chinese car users preferred the built-in navigation system, whereas this percentage increased to 54% in 2021. However, the increasingly rich functions, larger screens, and more complex information of automotive HMI systems may also lead to more severe driver distractions. Therefore, it is important to evaluate and optimize the safety of automotive HMI. If an HMI system can significantly reduce driver distraction, driving safety will undoubtedly be enhanced.

6.1.2 Types and Effects of Driver Distraction

A driver's attention is limited, and attempting to perform any secondary task can contribute to a decrease in driving performance due to distraction [10]. There are three types of driver distractions caused by secondary tasks: (1) visual distraction, where the driver diverts their gaze away from the road to interact with a device, leading to observation errors; (2) cognitive distraction, where the driver shifts their attention from driving to a secondary task, which can induce errors in information processing or memory retrieval; and (3) manual distraction, which occurs when the driver removes their hands from the steering wheel to operate other equipment, consequently leading to errors in physical actions [11, 12]. The causes and manifestations of these three types of distraction are summarized in Table 6.1.

Visual distraction weakens perception and increases fixation time; moreover, it is accompanied by considerable lateral lane departure. Additionally, it increases driver alertness, causing them to reduce speed and increase the following distance to compensate for their slower response to potential emergencies [13, 14]. Cognitive distraction affects the action predictions of other vehicles on the road. Many studies

Table 6.1 Causes and manifestations of the three main types of driver distraction

	Visual distraction	Cognitive distraction	Manual distraction
Causes	Gaze diversion from the road to observe the displays or buttons	Attention shift from driving to secondary tasks	Taking one hand off the steering wheel to operate other devices
Manifestations	Inability to observe the surrounding road environment	Slower processing and memory retrieval of driving-related information	Decreased control precision of the steering wheel
Major effects on driving	Reduced speed, lane departure, slower emergency response	Slower emergency response, increased following distance, going the wrong way	Lane departure

have found that cognitive distraction reduces gaze diversion and lane departure in simple road conditions without vehicles or sudden events [15, 16]. However, there is still controversy regarding whether cognitive distraction improves lane-keeping performance. Some studies suggest that time-to-line crossing is a more effective index than lane departure, as it indicates poorer lane-keeping safety during cognitive distraction [17]. Visual distraction during driving generally has a greater impact on driving safety than cognitive distraction, while perceptual impairments caused by visual distraction result in slower driver reaction than cognitive distraction [18]. Manual distraction primarily impairs the vehicle's lateral control ability and can be exacerbated by more complex body movements [19]. Despite the relatively small number of academic studies on manual distraction, its safety hazards are evident. During actual driving, when one hand leaves the steering wheel to operate the touchscreen or buttons, only the other hand holds the steering wheel, which may reduce control precision. Additionally, if the driver twists or tilts their body, this may cause hand tremors when gripping the steering wheel.

These three types of driver distractions exert combined effects, and hence, it is difficult to measure each type independently. Therefore, indexes from two categories, namely, driving performance and visual demand, are generally selected when conducting an objective evaluation of driver distraction. Measurement indexes for driving performance include longitudinal speed control, following distance control and brake reaction time as well as lateral steering errors, lane departure and time-to-line crossing. Driving performance issues can be caused by one or multiple types of distraction (i.e., visual, cognitive, or manual). Poor driving performance can directly lead to traffic accidents. For example, a short following distance or slow brake reaction can result in a rear-end collision, while lane departure may lead to crashes with oncoming vehicles or road guardrails. Measurement indexes for visual demand include glance counts, average glance duration, maximum glance duration, percentage of gaze diversion time, and total gaze diversion time [20–23]. Visual demand primarily addresses visual distraction and is minimally affected by cognitive and manual distraction. Excessive visual demand is one of the reasons for poor driving performance. However, driving performance evaluation cannot replace visual demand for two reasons. First, there is some randomness as to whether excessive visual demand leads to a significant decrease in driving performance, and analyzing only the latter may overlook potential safety hazards. Second, gaze track is one of the most fundamental data sources for understanding the causes of driver distraction, which also provides precise and detailed information. Thus, analyzing gaze track can facilitate the identification of issues in automotive HMI design and the formulation of targeted improvements.

6.2 Evaluation Indexes

In automotive HMI evaluation, safety can be divided into second-level evaluation indexes, including driving performance maintenance, emergency response, gaze diversion, and function restrictions.

6.2.1 Driving Performance Maintenance

Driving performance maintenance refers to the driver's ability to maintain a comparable level of driving performance while engaging in automotive HMI tasks, as they would without any interaction tasks. It is an integral aspect of overall driving performance that serves as a comprehensive assessment of visual, cognitive, and manual distractions. The index encompasses two crucial aspects of driving performance: longitudinal speed maintenance and lateral lane keeping.

Speed Maintenance

In everyday driving, vehicles spend most of the time maintaining a constant speed. On roads with low traffic volume, this speed is generally the speed limit, while on roads with higher traffic volume, it is generally lower than the speed limit and is the common speed of surrounding vehicles. When driving at a constant speed, the driver's primary task load is low, which makes it a suitable time to use the automotive HMI system. Drivers usually avoid using the HMI system during acceleration and braking.

Speed maintenance ability is reflected in the magnitude of speed deviation, that is, the average deviation between the actual and target speeds, which is given by

$$SpDev = \sum_{t=0}^{T} |v_t - v_0|,$$

where, $SpDev$ is the speed deviation, v_t is the actual speed at time t, v_0 is the target speed, and T denotes the impact time of the interaction task on driving performance maintenance. The unit of T corresponds to the sampling interval of the actual speed. Notably, the absolute value of the difference between the actual and target speeds must be taken to prevent positive and negative deviations from cancelling each other out. The impact time of the interaction task on driving performance maintenance is longer than that of the interaction task itself. It starts at the beginning and extends beyond the end of the task, which, according to experience, can be set to 3 s after the task ends.

Figure 6.1 shows the actual speed variation curve of a certain car model during a test with $v_0 = 60$ km/h. In the absence of interaction tasks, the driver could maintain the speed very well at 60 km/h. However, when using the touchscreen to switch to the next song-track, the driver was unable to continuously observe the speed information due to visual distraction and was also incapable of precisely controlling the accelerator pedal owing to cognitive distraction, resulting in an elevated speed. After 2.5 s into the task, the driver became aware of the speed deviation and started to decelerate. However, even though the task was completed in 5.4 s and the driver was no longer visually distracted, their attention did not return to the driving task instantly and completely; thus, the speed continued to decrease at a constant rate. It was not until approximately 8.5 s later that the driver started adjusting the accelerator pedal to gradually restore the target speed. Therefore, it is still necessary to record data within 3 s after task completion. For simple one-step tasks with a short duration, it is particularly important to extend the recording time to 3 s after the task ends because insufficient speed deviation accumulates within the brief task duration, and peak deviation often occurs after task completion.

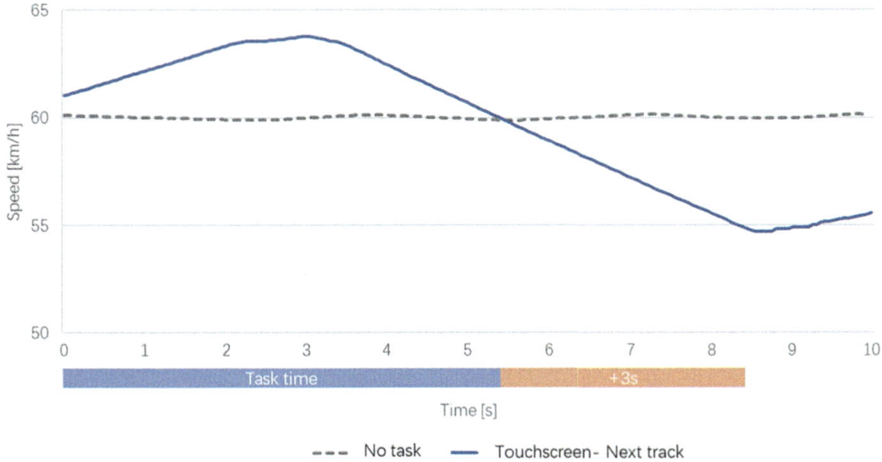

Fig. 6.1 Speed variation curve of a certain car model in the absence of interaction tasks, and when the driver is switching to the next song-track using the touchscreen

Lane Keeping

While driving on roads with lane markings, vehicles are expected to remain within a designated lane, following its path and making turns in synchrony with the lane, unless the driver consciously decides to change lanes. When driving along a lane, the driver's primary task load is relatively low, which makes it a suitable time to use the automotive HMI system. Drivers usually avoid using the HMI system when changing lanes or crossing intersections.

Lane-keeping ability is reflected in the standard deviation of lane departure, which corresponds to the standard deviation of the vehicle's actual lateral coordinates, which is given by

$$LDSD = \sqrt{\frac{1}{T}\sum_{t=0}^{T}(d_t - d_{avg})^2},$$

where $LDSD$ is the standard deviation of lane departure, d_t is the actual lateral coordinate of the vehicle relative to the lane centerline at time t, d_{avg} is the average lateral coordinate of the vehicle within time period T, and T denotes the impact time of the interaction task on driving performance maintenance. The unit of T corresponds to the sampling interval of the actual speed. In lane keeping, the standard deviation of the lateral position rather than the absolute deviation relative to the lane centerline should be considered. As the driver sits on the left side of the vehicle (or on the right side for right-hand drives) rather than the lateral center, it is difficult to accurately judge the precise lateral position of the vehicle and keep it exactly on the absolute centerline of the road. As long as there are no sudden, unexpected lateral position changes, this will not pose a potential hazard to driving safety even if the vehicle travels slightly to the left or right within the lane. Once the vehicle crosses the lane markings, it may collide with adjacent vehicles or road guardrails. However, automotive HMI systems typically do not cause significant distractions in drivers, which implies it is unnecessary to consider such situations during the evaluation of lane-keeping ability. In addition, as with speed maintenance, it is important to account for the extended impact time of the interaction task on driving performance maintenance when assessing lane-keeping performance, which persists for approximately 3 s following task completion.

Figure 6.2 shows the actual lateral position variation curve of a certain car model during a test. In the absence of interaction tasks, the driver could maintain a relatively straight path at an approximate position of 0.12 m. However, when engaged in the task of switching to the next song-track using the touchscreen, the driver was unable to continuously monitor their position due to visual distraction. Additionally, cognitive distraction hindered the precise control of the steering wheel angle, while manual distraction led to hand tremors when gripping the steering wheel. These factors collectively contributed to a constant deviation in the vehicle's lateral position. It was only after the task concluded at 5.4 s did the driver become aware of the lane departure and subsequently initiated corrective action by turning the steering wheel in

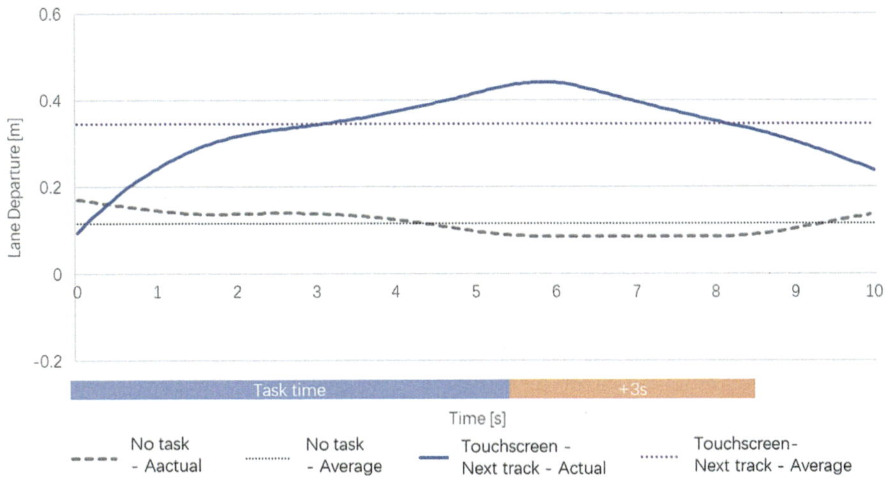

Fig. 6.2 Lateral position variation curve of a certain car model in the absence of tasks, and when the driver is switching to the next song-track using the touchscreen

the opposite direction, eventually returning the vehicle to the average lateral position at 8 s.

6.2.2 Emergency Response

Safe driving is more than maintaining a constant speed along a fixed lane. Drivers may need to apply the brake or make sudden turns to avoid accidents in emergencies, such as when pedestrians unexpectedly cross the road, surrounding vehicles abruptly change lanes, or sudden obstacles appear on the road. As a part of driving performance maintenance, emergency response is a comprehensive assessment of visual and cognitive distraction, as well as the most intuitive approach to examining cognitive distraction. Emergency response assessment focuses on the driver's response in scenarios requiring emergency braking. Although emergency steering can also prevent accidents, emergency braking is generally advocated, in accordance with the principle of "reducing speed over changing lanes". Moreover, many countries' traffic laws tend to penalize unreasonable lane changes when determining liability for traffic accidents.

Emergency response can be assessed through the driver's braking reaction time, that is, the interval between the occurrence of an emergency and the moment when the driver presses the brake pedal. It can also be reflected in time-to-collision (TTC), which refers to the time it would take for a collision to occur if the test vehicle and the target vehicle in front continue to travel at the current speed when an emergency occurs (e.g. at the time of abrupt brake application by the vehicle in front). Assuming that no collision should occur, the shorter the minimum TTC at the time of the emergency, the faster is the driver's reaction speed, which indicates that the driver was less affected by distractions. TTC is commonly used in the testing of automatic emergency braking (AEB) and forward collision warning (FCW) systems [24].

Although emergency response while operating the automotive HMI system is crucial, accurately measuring it in real-world scenarios poses a considerable challenge. First, whether using real road or driving simulation tests, it is difficult to precisely control the trigger time of emergencies and achieve uniformity in the different forms of emergencies (especially in the case of pedestrians crossing the road). More importantly, the degree of cognitive distraction fluctuates when drivers operate HMI systems, and periods of visual distractions are interspersed throughout the task. Thus, even slight differences in the trigger time of emergencies can lead to substantial variations in driver reaction times. Taking visual distraction as an example, when the vehicle in front suddenly applies the brakes, the reaction will be longer if the driver is gazing at the central information display. Conversely, if the driver is in the interval between two gazes at the central information display and is concentrating on the road, the reaction time will be shorter. The driver's gaze shifts are extremely fast and difficult to predict in advance. During the testing process, it is challenging to trigger an event only when the driver's gaze is within a specific fixed area. Therefore, the evaluation of emergency response usually requires a large sample size to eliminate biases. While this is feasible in scientific experiments, it is inefficient in actual product testing, and hence this index is sometimes disregarded.

6.2.3 Gaze Diversion

Gaze diversion, which is the most direct test of visual distraction, refers to the shift of the driver's gaze from the road in front to the screen or button area when operating the automotive HMI system.

When evaluating gaze diversion, it is necessary to divide the driver's FOV into different areas of interest (AOIs). Typically, two fundamental AOIs can be established, as depicted in Fig. 6.3. The first AOI encompasses the road scenario visible through the windshield (indicated by the orange area), while the second AOI comprises the central information display, instrument cluster display, and button area (indicated by the green area). When the driver operates the HMI system, their gaze inevitably transitions from the road scenario to the screen or button area. However, for the sake of driving safety, they intermittently redirect their gaze back to the road, ensuring no driving performance risk, before resuming their gaze on the screen or button area again. Therefore, the driver's gaze will repeatedly shift back and forth between these two AOIs, and almost all the fixations will settle within them, as shown in Fig. 6.4. For driving or HMI tasks that require checking the rear-view mirror, the position of the rear-view mirror is sometimes treated as a separate AOI. When conducting more fine-grained gaze analysis, the AOIs can be further subdivided. For example, the central information display can be an AOI, and one of its icons can be a smaller AOI. However, such fine-grained AOI division is not essential when evaluating gaze diversion.

The smallest unit of eye gaze is a fixation, that is, the brief dwelling of a gaze shift within a given time period. This dwelling allows the eyes to focus on a specific point within an AOI and causes the point to fall on the fovea centralis on the retina. Each green dot in Fig. 6.4 represents a fixation, with the duration ranging from 100 to 2,000 ms. Saccades are brief and rapid eye movements between adjacent fixations.

Fig. 6.3 Division of the two basic AOIs in front of the driver

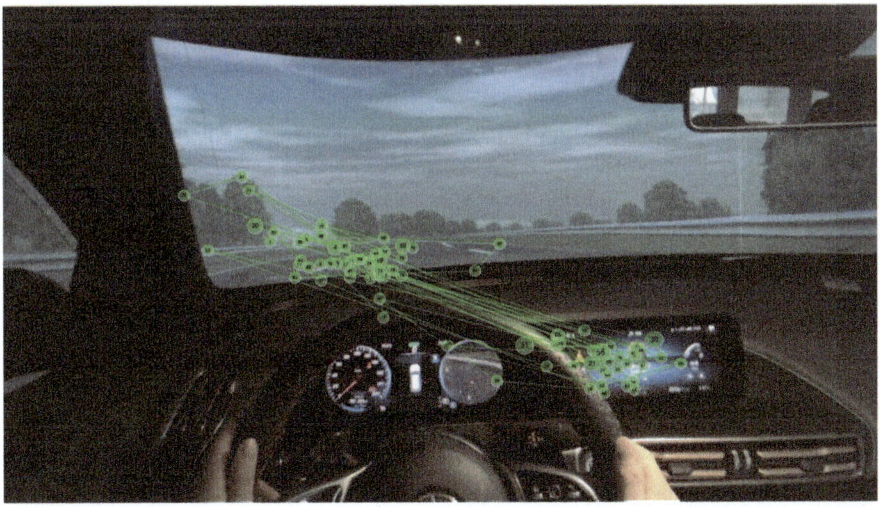

Fig. 6.4 Examples of the driver's gaze track when using the central information display

The line connecting two adjacent green dots in Fig. 6.4 represents one saccade. The maximum saccade speed can reach up to 500°/s, and its distance usually ranges from 1° to 5°. A glance refers to the dwelling of eye gaze within a specific AOI, which generally involves multiple fixations and saccades. Glance duration includes the duration of all fixations within the AOI, blink duration, saccade duration between fixations or blinks, as well as the duration of one saccade prior to the first fixation within the AOI, that is, the transition time from the previous AOI to the current one. Typical glance durations range from 500 to 3,000 ms. If the saccade duration before the first fixation is excluded, the time that the gaze remains within the AOI is referred to as visit duration or dwell time.

The two most important evaluation indexes in gaze diversion are the total and maximum glance durations. The AOIs for glances consist of the central information display, instrument cluster display, buttons, and their surrounding areas, as represented by the green area in Fig. 6.3. The total glance duration is the total duration of gaze diversion caused by a specific interaction task. The longer the total duration, the longer the driver's attention is diverted from observing the road, posing a greater safety hazard. Glances caused by interaction tasks are typically intermittent rather than continuous; therefore, each individual glance must be examined. The maximum glance duration refers to the longest continuous duration of gaze diversion caused by a specific interaction task. During a single glance, the driver cannot observe their driving conditions, which causes increased lane departures and an inability to monitor changes in their surroundings, thus making it difficult to respond to emergencies. Therefore, two times of 3 s glances are more dangerous than three times of 2 s glances, despite a total glance duration of 6 s. When the total glance duration is constant, driving safety can be enhanced by minimizing the duration of each individual glance, especially that of the longest glance. When it is difficult to measure the maximum glance duration, the average glance duration can be considered. However, it is important to note that the measurement metrics for the average duration should be stricter than those for the maximum duration.

The Alliance of Automobile Manufacturers (AAM) recommends that the total glance duration required to complete an interaction task should not exceed 20 s, with each individual glance lasting less than 2 s [25], that is, the maximum duration of a single glance should not exceed 2 s. The AAM's recommendation is relatively lenient. At a speed of 72 km/h, a 2-s glance would prevent the driver from observing the road for a distance of up to 40 m. Therefore, when designing an automotive HMI system, efforts to minimize visual distraction should not be limited to meeting AAM's standards alone.

6.2.4 *Function Restrictions*

With the continuous advancement of automotive HMI systems, an increasing number of entertainment and lifestyle services unrelated to driving are being integrated into vehicles. Many of these functions require the driver to fixate on the screen for a long time, causing significant visual distraction; therefore, these services should be restricted during driving. For example, playing movies inside the vehicle inevitably requires users to fixate on the screen for an extended period of time. Even without measuring gaze diversion, we know that playing movies while driving poses a considerable safety hazard.

Function restrictions involve filtering out functions that are unrelated to driving and pose significant risks of visual distraction. During driving, the central information and instrument cluster displays should not present functions that require the driver's fixation for an extended period of time. This includes, but is not limited to, watching videos, playing games with dynamic graphics, and reading long, unsorted lists (e.g. restaurant menus, news, and microblogs). Not all lists should be prohibited, as some lists only cause limited visual distraction and may be necessary while driving. Lists with priority sorting should not be prohibited. For example, after the driver enters text into the navigation system and a list of relevant destinations is displayed, the driver typically selects one of the top three options, which does not require prolonged fixation. Short lists within 2–3 pages, such as vehicle setting menus or song lists of limited length recommended by the system, should also not be prohibited.

6.3 Summary of Evaluation Results and Design Suggestions

This section will present an analysis of the safety level for each typical interaction modality and provide suggestions for design optimization [26]. Recommendations for selecting interaction modalities for specific tasks are also discussed in detail in Chap. 7. When selecting the optimal interaction modality for each interaction task, it is necessary to consider both safety and efficiency. Furthermore, the safety evaluation results often correlate with efficiency results to some extent.

6.3.1 Central Information Display

The touchable central information display (touchscreen) is currently the core modality of automotive HMI and will remain irreplaceable in the near future. With its relatively high information transmission efficiency, the touchscreen can display as many as 20–40 icons and phrases of different levels on a single page, and the driver can locate their desired target on this page within 0.5–2 s. This efficiency surpasses that of physical buttons and voice control. However, higher information transmission efficiency also increases the risk of driver distraction, which implies that the touchscreen is a mainstream interaction modality with poor safety.

In a comprehensive HMI usability test conducted by the authors in 2019 involving eight mainstream intelligent vehicles in the Chinese market, we found that touchscreens had significant safety disadvantages compared to voice control, central console buttons, and steering wheel buttons. First, touchscreens had an adverse impact on speed maintenance, resulting in a 14% higher speed deviation compared to the averages of the other three interaction modalities. Second, similar to voice control, touchscreens affected lane keeping, with a 52% higher standard deviation of lane departure compared to the averages of the other two interaction modalities. Third, touchscreens almost definitely led to gaze diversion, with a possibility of less than 5% for blind operations (including peripheral vision operations). Fourth, touchscreens had the longest total glance duration, which was 191% higher than the averages of the other three interaction modalities. Finally, the average glance duration for touchscreens was significantly higher (93% higher) than the averages of the other three interaction modalities.

During the design stage, driver distraction resulting from touchscreens can be mitigated and safety can be enhanced by optimizing the hardware and software of touchscreen interaction. Based on the random forest algorithm, the importance of the optimizable variables in touchscreen interaction design for each safety index is shown in Table 6.2. Variables with an importance value of less than 0.07 are considered unimportant and are not included in the analysis.

The top five most important variables are discussed as follows.

Table 6.2 Importance values of random forest independent variables in touchscreen interaction design (importance value > 0.07)

Independent variables	Dependent variables (evaluation indexes)			
	Speed deviation	Standard deviation of lane departure	Total glance duration	Average glance duration
Spherical surface area				0.122
Distance to screen center	0.127			
Horizontal rotation angle				0.154
Operation steps		0.161	0.232	
Rightmost position of the operation area	0.092	0.081		0.078
Topmost position of the operation area	0.089			
Click Displacement		0.125	0.174	
Average touchpoint area	0.078			0.146
Percentage of box-type touchpoints		0.089		

(i) **Operation steps:** This is also a third-level efficiency index. Operation steps have a considerable impact on lane departure and total glance duration. A one-step operation results in minimum values for both the lane departure and total glance duration. However, one-step operation tasks are very rare in actual interaction design. A one-step operation means that the task icon is directly placed on the homepage of the central information display; however, due to the limited homepage area, it cannot accommodate a large number of task icons. For many models, the homepage typically has only the menu icons for first-level directories, without any specific task icons. Furthermore, the advantage of central information displays over physical buttons is that the positions of their icons can be changed to achieve the "folding" of interface. Therefore, even for basic tasks that are most frequently used, such as adjusting the air-conditioning temperature or volume, the optimal design is still physical buttons rather than a quick one-step icon on the homepage of the central information display. Two-step tasks result in a significant increase in lane departure but a relatively limited increase in total glance duration. Therefore, for tasks that are relatively frequently used, it is preferable to design them as two-step operations. In general, a two-step task involves entering a second-level directory menu (e.g. the music and navigation menus) in the first step, and selecting the specific task icon in the second step. This requires a simple and efficient logical structure design for the menu, and frequently used functions can be placed in second-level directory menus for quick access, thus eliminating the need to go down level by level.

(ii) **Click Displacement:** This refers to the total distance of finger movement on the screen from the first step to the last step when users use the touchscreen to operate a function. Click displacement is comparable to operational displacement in the third-level indexes of efficiency, but it excludes the distance covered from the moment the finger enters the screen area until the first tap. This variable has a significant impact on lane departure and total glance duration. When click displacement reaches 20 mm, the total glance duration increases significantly, and when it reaches 60 mm, lane departure starts to increase rapidly. For tasks requiring two or three steps, it is possible to design the icons of each step to be in the same or similar position as much as possible, thereby keeping the finger movement within 60 mm or even 20 mm. This requires designer of central information display interactions to have a strong predictive ability for the driver's operation steps, based on which targeted designs can be incorporated. For tasks that involve four or more steps, it is very difficult to keep the finger movement distance within 60 mm. In such cases, it should be kept within 240 mm to avoid excessive lane departure. As shown in Fig. 6.5, in the task of calling a designated contact, the icons of the first and second steps are close to each other for ease of operation, while the icon of the fourth step is far away from those of the third and fifth steps, resulting in a larger click displacement.

(iii) **Rightmost Position of the Operation Area:** This refers to the lateral angle θ corresponding to the center position of the rightmost icon in spherical coordinates across all steps of a given interaction task. The measurement method

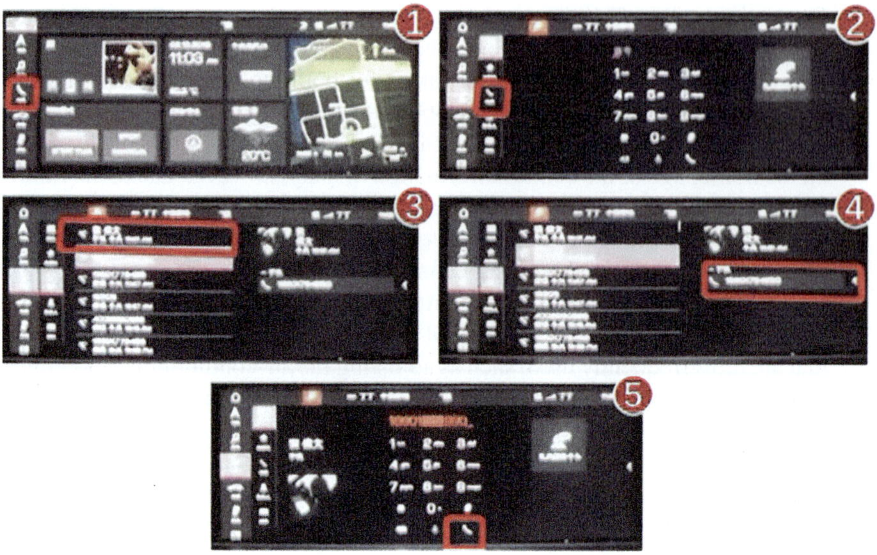

Fig. 6.5 Images of the five steps required to call a designated contact using the touchscreen of the Porsche Panamera (2018 model)

for the lateral angle in spherical coordinates is shown in Fig. 6.6. The rightmost position of the operation area significantly affects the three evaluation indexes: vehicle speed deviation, lane departure, and average glance duration. The optimal distribution of the rightmost position is between 35° and 45°. An excessively large angle will make it difficult for the driver's arm to reach positions that are too far to the right without body movement, which can result in unstable left hand movements of the driver, which controls the steering wheel. As a result, positions beyond 47° can cause a considerable increase in lane departure. Conversely, an overly large angle will also require drivers to shift their gaze substantially to the right, and this extended fixation path can increase the duration of each glance. The average glance duration gradually increases when the rightmost position of the operation area exceeds 45°. As shown in Fig. 6.5, in the task of calling a designated contact, the fourth step involves the rightmost icon, which has a lateral angle exceeding 45°, thereby affecting safety. When the rightmost position of the operation area is less than 35°, it can also pose challenges for observation. This goes against common judgement, as the proximity of the central information display to the driver does not necessarily make it easier to perceive the displayed content. The increase in average glance duration caused by a rightmost position of less than 35° may be attributed to the fact that on the left side of the central information display, the driver's gaze also includes the right end of the steering wheel and the right part of the instrument cluster display. Moreover, some car models may have more complex dashboard or air vent designs that are often located on different

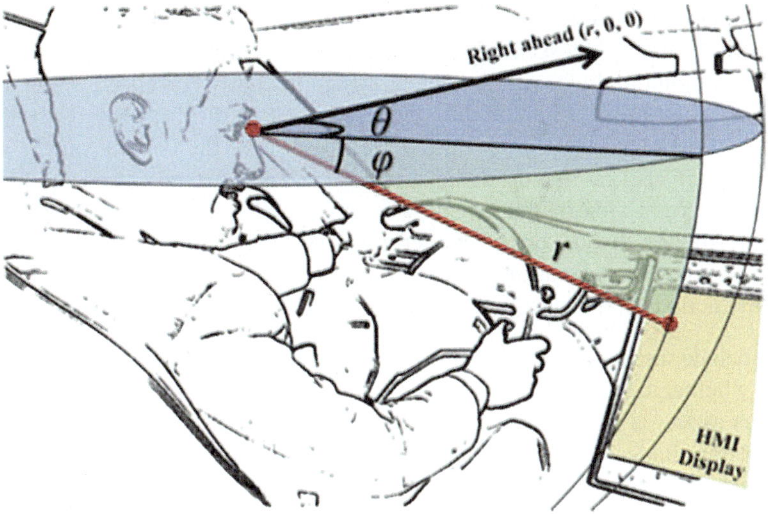

Fig. 6.6 Measurement of spherical coordinates of the central information display centered on the driver's eyes

planes, thus creating a layered and intricate visual environment. Even if these elements do not directly obstruct the content on the left part of the central information display, the complex surrounding environment can still have an impact on the driver's focus, and there may be significant differences in the images received between the two eyes. In contrast, the right side of the central information display is typically surrounded only by the dashboard. The elements in this area are relatively simple, making it easier for the eyes to focus, and thus there are no significant differences in binocular vision.

(iv) **Average Touchpoint Area:** This refers to the average area of all icons that need to be tapped for a particular interaction task. This variable is important for the average glance duration but also has some impact on vehicle speed deviation. For average glance duration, 700 mm^2 is a significant threshold. When the touchpoint area is less than this threshold, the average glance duration gradually decreases with an increase in icon size. However, when the touchpoint area exceeds 700 mm^2, it no longer has a significant impact on the average glance duration. As drivers need to control their entire arm through their shoulder when operating the central information display, the force arm is far longer than that when operating a cell phone. Therefore, larger icons on the central information display make it easier for drivers to locate and tap. For the most frequently used icons, 700 mm^2 is a suitable size, e.g., 35 $\times$ 20 mm. Larger icons would compromise the aesthetic appeal and flexibility of the interface design without significantly reducing visual distractions. For vehicles that frequently travel on bumpy roads, such as off-road vehicles, larger icons can be considered.

(v) **Horizontal Rotation Angle:** This refers to the angle between the projected vertical line of the central information display on the horizontal plane and the longitudinal axis of the vehicle, which is highly correlated to the average glance duration. Orienting the central information display towards the driver's side will make it easier for the driver to observe the content on the screen. This will also shorten the distance the driver's arm extends when tapping icons on the right side of the screen. The horizontal rotation angle of the central information display does not need to be perfectly square to the driver, which is impossible. A rotation angle of 7° is sufficient to significantly reduce visual distractions and is easily achievable for most automotive interior designs.

The debate surrounding the layout of the central information display often revolves around whether it should be horizontal or vertical. However, in terms of driver distraction, this aspect is not a significant concern. The aspect ratio of the display holds minimal importance in the random forest calculation for safety evaluation indexes and can be considered negligible. In contrast, the area occupied by the central information display in spherical coordinates plays a more crucial role, particularly in determining the average glance duration. A larger display size results in a longer average glance duration as it becomes more challenging for the driver to locate specific icons within a larger area. By appropriately dividing a larger central information display into different regions, it is possible to reduce the average glance duration.

6.3.2 Buttons

When analyzing the impact of buttons on driver distraction, it is necessary to distinguish between function and directional buttons, as shown in Fig. 6.7, both of which can be found on the steering wheel or in the central console area. Function buttons serve a specific and predetermined purpose, such as returning to the homepage or controlling the volume, and can complete the task in a single step. Users typically do not need to rely on the information shown on the central information display or instrument cluster display to determine how to use these buttons. Function buttons can take the form of a regular button, a scroll wheel, or one of the directions on a joystick.

By contrast, directional buttons do not have a fixed function and have to be operated in conjunction with the information shown on the central information display or instrument cluster display. For example, they can be used to navigate the cursor on the screen to go up, down, confirm, or return. Directional buttons can take various forms, including scroll wheels, joysticks, knobs, touchpads, or regular buttons.

Operating function buttons cause minimal driver distraction and rarely affects safety. Compared to icons on a touchscreen, function buttons have a completely fixed position, which makes it easier for drivers to remember and locate them, thereby reducing cognitive load and gaze diversion. Additionally, most types of buttons (except for non-vibrating touch buttons) provide distinct tactile feedback, which

Fig. 6.7 Function and directional buttons on the steering wheel and central console area of the Mercedes-Benz C-Class (2019) (*Source* The Mercedes-Benz Group)

helps the driver confirm the location of the button press and whether the button press was successful, thus minimizing gaze diversion. Naturally, if the buttons are positioned too close to each other, the driver may find it more difficult to locate them. However, such cases are rare in the current trend of using fewer physical buttons. Function buttons on the steering wheel offer better safety compared to those in the central console area because, when operating steering wheel buttons, the driver only needs to move their thumb without taking their palm off the wheel, thus causing minimal manual distraction. Moreover, using only the thumb for movement makes it easier to locate the target button by touch, enabling the possibility of blind operation, which implies less visual distraction.

Directional buttons cause greater driver distraction compared to function buttons, even greater than using the touchscreen. Directional buttons must be operated in conjunction with the information displayed on the screen, which means that the driver must fixate on the screen for a long period of time, similar to using touchscreen interactions, thus losing one of the major advantages of physical buttons. Furthermore, the process of using directional buttons is often more cumbersome. For example, when we need to tap on the fifth icon on the screen, a touchscreen allows direct interaction with a finger; however, when using directional buttons, the cursor needs to be moved four steps to reach the fifth icon. Therefore, for touchable central information displays, directional buttons have limited value, whereas for non-touchable instrument cluster displays, directional buttons still play an important role in controlling the menus.

Directional buttons on central information displays have a unique development history. Two decades ago, touchscreen technology was not advanced, and central information displays at that time had smaller sizes and linear logic structures. Many car manufacturers designed their own directional buttons to control non-touchable central information displays, and produced excellent user experiences, as exemplified by BMW's early iDrive and Mercedes-Benz's Command systems. However, as technology and design concepts evolved, relying on directional buttons for central information display interactions was no longer advantageous. Nevertheless, keeping directional buttons as an additional option might be beneficial for some users, as long as the experience of other interaction modalities is not affected.

6.3.3 Voice Control

Among all interaction modalities, voice control causes the least visual distraction. According to the test results of eight mainstream intelligent vehicles in the Chinese market by the authors in 2019, approximately 40% of voice interaction tasks can be performed without gaze diversion, i.e., blind operation. For simple tasks, the percentage of blind operations can increase to approximately 55%. In theory, voice interaction should achieve a 100% blind operation rate; however, this is not the case in practice. Many users are accustomed to looking at a specific image when using voice control, such as an anthropomorphic virtual assistant or an abstract dynamic image. This habit stems from our tendency to look into the eyes of the person we are speaking to. Although this gaze itself has no practical significance for the interaction, it is a habit that people find difficult to change. On the other hand, for some interaction tasks, the central information display itself is an important output device for voice control. Reading the information displayed on the screen can improve interaction efficiency and enhance user trust. For example, it is more efficient to display a list of alternatives on the screen after inputting a navigation destination using voice control than to read them out sequentially. Furthermore, when the user gives a command and the interaction system does not respond immediately with sound, the user can judge from the screen display whether the system is processing the command or if the system did not hear it.

For more complex tasks, such as deep hierarchical tasks or tasks requiring specific text input, voice control usually involves fewer steps and is faster than other interaction modalities. This shortens the total time for various types of driver distractions, which can effectively avoid the continuous accumulation of speed deviation and lane departure, as well as reduce gaze diversion.

However, there are limitations to voice control, and it can sometimes cause greater cognitive distraction than other interaction modalities. While conversing with a real person in the vehicle can divert the driver's attention and cause cognitive distraction, the cognitive distraction resulting from interacting with an HMI system can be even more significant. First, voice control requires specific command phrases that the driver needs to consciously recall and say without prompts, such as "Set the temperature to X degrees." Despite the diverse and naturalistic commands recognized by many car models, current voice recognition systems still fall short of human comprehension, making it difficult for drivers to talk with them naturally without conscious effort. In comparison, adjusting the air-conditioning temperature with a button only requires the driver to instinctively press a button in a fixed position, which requires minimal cognitive load. Additionally, when providing longer voice commands, drivers must speak without interruptions, word repetitions, or interjections such as "uh" or "um" to ensure accurate recognition. This requirement further demands heightened concentration during speech.

While various forms of driver distraction caused by different interaction modalities are directly related to task complexity, voice control exhibits a relatively high lower bound for cognitive distraction. Therefore, for simpler interaction tasks, voice control may not fully demonstrate its advantages in reducing visual distraction but instead will further highlight its disadvantages in increasing cognitive distraction.

References

1. Yao Y, Zhao X, Du H, et al. Classification of distracted driving based on visual features and behavior data using a random forest method [J]. Transportation Research Record, 2018, 2672(45): 210–221.
2. Née M, Contrand B, Orriols L, et al. Road safety and distraction, results from a responsibility case-control study among a sample of road users interviewed at the emergency room [J]. Accident analysis and prevention, 2019, 122: 19–24.
3. Stutts J C, Reinfurt D W, Staplin L, et al. The role of driver distraction in traffic crashes [R]. Chapel Hill: University of North Caroline Highway Safety Research Center, 2001.
4. Regan M, Victor T, Lee J. Driver Distraction and Inattention: Advances in Research and Countermeasures Volume I [M], Taylor & Francis, 2013.
5. McEvoy S P, Stevenson M R, Woodward M. The impact of driver distraction on road safety: results from a representative survey in two Australian states [J]. Injury Prevention: Journal of the International Society for Child and Adolescent Injury Prevention, 2006, 12(4): 242–247.
6. Huemer A K, Schumacher M, Mennecke M, et al. Systematic review of observational studies on secondary task engagement while driving [J]. Accident Analysis and Prevention, 2018, 119(5): 225–236.
7. Metz B, Landau A, Just M. Frequency of secondary tasks in driving - Results from naturalistic driving data [J]. Safety Science, 2014, 68, 195–203.
8. Sayer J R, Devonshire J M, Flannagan C A. The Effects of Secondary Tasks on Naturalistic Driving Performance [R]. Ann Arbor, Michigan: The University of Michigan, 2005.
9. Crundall E, Large D R, Burnett G. A driving simulator study to explore the effects of text size on the visual demand of in-vehicle displays [J]. Displays, 2016, 43, 23–29.
10. Choudhary P. Mobile phone use during driving: effects on speed reduction and effectiveness of compensatory behaviour [J]. Accident Analysis and Prevention, 2017, 106 (6): 370–378.
11. Strayer D L, Watson J M, Drews F A. Cognitive distraction while multitasking in the automobile [J]. Psychology of Learning and Motivation - Advances in Research and Theory, 2011, 54(54): 29–58.
12. Young K L, Salmon P M. Examining the relationship between driver distraction and driving errors: A discussion of theory, studies and methods [J]. Safety Science, 2012, 50 (2): 165–174.
13. Muhrer E, Vollrath M. The effect of visual and cognitive distraction on driver's anticipation in a simulated car following scenario [J]. Transportation Research Part F: Psychology and Behaviour, 2011, 14, 555–566.
14. Kountouriotis G K, Spyridakos P, Carsten O M J, et al. Identifying cognitive distraction using steering wheel reversal rates [J]. Accident Analysis and Prevention, 2016, 96, 39–45.
15. Engström J, Markkula G, Victor T, et al. Effects of Cognitive Load on Driving Performance: The Cognitive Control Hypothesis [J]. Human Factors, 2017, 59(5): 734–764.
16. Metz B, Schoch S, Just M, et al. How do drivers interact with navigation systems in real life conditions? [J]. Transportation Research Part F: Psychology and Behaviour, 2014, 24, 146–157.

17. Li P, Merat N, Zheng Z, et al. Does cognitive distraction improve or degrade lane keeping performance? Analysis of time-to-line crossing safety margins [J]. Transportation Research Part F: Psychology and Behaviour, 2018, 57, 48–58.
18. Jin L, Xian H, Niu Q, et al. Research on safety evaluation model for in-vehicle secondary task driving [J]. Accident Analysis and Prevention, 2015, 81, 243–250.
19. Libby D, Chaparro A, He J. Distracted while driving: a comparison of the effects of texting and talking on a cell phone [J]. Proceedings of the Human Factors and Ergonomics Society Annual Meeting, 2013, 57(1): 1874–1878.
20. Hofmann H, Tobisch V, Ehrlich U, et al. Evaluation of speech-based HMI concepts for information exchange tasks: A driving simulator study [J]. Computer Speech Language, 2015 33(1): 109–135.
21. Liang Y, Lee J D. Combining cognitive and visual distraction: Less than the sum of its parts [J]. Accident Analysis and Prevention, 2010, 42(3): 881–890.
22. Gaspar J G, City I, Ward N, et al. Measuring the useful field of view during simulated driving with gaze-contingent displays [J]. Human Factors, 2016, 58(4): 630–641.
23. Rosenthal T J. STISIM Drive User's Manual [Z]. Systems Technology Inc, 1999.
24. M Minderhoud, P Bovy. Extended time-to-collision measures for road traffic safety assessment [J]. Accident Analysis and Prevention, 2001, 33, 89–97.
25. AAM. Statement of Principles, Criteria and Verification Procedures on Driver Interactions with Advanced In-Vehicle Information and Communication Systems [R]. (2006-06-26).
26. Gong Zaiyan. In-vehicle HMI Driving Distraction Evaluation and Optimization Design Method [D]. Shanghai: Tongji University, 2020.

Chapter 7
Efficiency

7.1 Development

Efficiency, one of the three basic usability indexes defined in the ISO-9241 international standard, refers to the resources consumed by the user and the HMI system for the completion of a specified interaction task. For users, the resources primarily include the task time, physical movements, and visual attention, whereas, for HMI systems, the resources primarily include the response time.

Efficiency is a vital objective in the working of all machines. The invention of the steam engine reduced the consumption of human and animal labor resources while also increasing the factory output. The advent of computers reduced dependence on human mental resources, and the accelerated development of computer systems increased computational speed. The invention of the Internet eliminated the need for physical media (such as letters) for information transmission, thus eliminating the transportation costs for physical media and enhancing the speed of information transmission. Similarly, automotive HMI systems aim to enhance the task execution speed of users in the vehicle and reduce the demand for their mental and physical resources.

Automotive HMI efficiency has two characteristics. First, it emphasizes the consumption of the user's resources rather than the resources of the HMI system. For software systems, most interaction tasks are simple and fast; hence, it is usually unnecessary to specifically quantify the electrical and computational power consumed. The system response time is sufficient to nearly reflect the computational resource consumption for each task. Additionally, the time required to complete an interaction task already includes the system response time. Therefore, for most tasks, the system response time can be considered a part of the user's resource consumption. Only tasks that do not require active user input, such as navigation system rerouting after a vehicle deviates from its planned route, will need a separate discussion on the system response time. Second, automotive HMI efficiency typically requires interaction tasks to serve as secondary tasks such that the driver is not fully engrossed in

J. Ma and Z. Gong, *Automotive Human-Machine Interaction (HMI) Evaluation Method*, Research on Automotive Intelligent Cockpit,
https://doi.org/10.1007/978-981-99-9951-4_7

operating the HMI system but instead prioritizes normal driving. Tasks that can only be performed when the vehicle is parked are relatively few and used less frequently. They typically do not focus on operational efficiency and thus sometimes can be excluded from the primary consideration of efficiency. Examples include options with deep logical hierarchies in the setting menu and watching video content.

Therefore, for automotive HMI systems, efficiency primarily refers to the ability of the system to improve the driver's operational efficiency and reduce their workload when performing tasks within the HMI system while driving the vehicle.

The ISO-9241 international standard provides a simple and general method for assessing the efficiency of interaction systems; however, it is not specifically designed for automobiles. In this method, the time spent by participants for completing each task is recorded. Each task begins with uttering the word "start" and ends when the user indicates completion. After each task is completed, users are required to answer the Single Ease Question (SEQ) survey to assess the task difficulty level. The SEQ survey consists of a single question and employs a 7-point Likert scale, where "1" indicates that the task is "very difficult" and "7" indicates that the task is "very easy." The average SEQ score for interaction systems across various domains is approximately 5.5.

The measurement and evaluation of automobile HMI efficiency can be further subdivided based on the task time and difficulty level mentioned in ISO-9241 for a more targeted investigation and to facilitate the formulation of a more standardized and objective operation.

7.2 Evaluation Indexes

In automotive HMI evaluation, efficiency can be divided into two second-level evaluation indexes: task time and operation complexity.

7.2.1 Task Time

Task time refers to the time taken from the start of an interaction task to the achievement of the final objective. The concept of task time is straightforward; however, in actual testing, the starting and ending points of the task must be clearly defined. Most tasks in automotive HMI systems need to be triggered by user actions, and the time taken to complete these tasks is referred to as the operation time. For tasks that do not require user operation, the response time is used as the metric.

Operation Time

Operation time refers to the time taken from the moment the driver starts executing an interaction task to the achievement of the final objective. The starting point of an interaction task could be the time point at which the driver receives the task instructions or spontaneously decides to execute the task, or it could be when they begin taking actions to perform the task. The time difference between the two moments represents the time taken by the driver to comprehend the task and think about how to perform the first step. For tasks that are familiar to the driver, this time difference is usually brief. Theoretically, choosing the former as the starting point is more logical; however, this is challenging in practice. During the process of receiving instructions, the driver might understand what the task entails after hearing the first few words without having to listen to the full instruction. This makes it difficult to determine the exact moment of comprehension. For tasks that a driver spontaneously decides to perform, the moment at which the decision is made is not overtly visible, making it even more challenging to measure. Therefore, in practice, the starting point of a task is generally taken as the time point at which the driver begins to perform actions. In terms of different interaction modalities, this can be determined as follows:

- Touchscreen: the time point at which the driver's hand leaves the steering wheel.
- Central console buttons: the time point at which the driver's hand leaves the steering wheel.
- Steering wheel buttons: the time point at which the driver's finger starts moving.
- Voice control: the time point at which the driver utters the voice assistant activating word or when the activating button is pressed (for models without a activating word).

The endpoint of an interaction task is the time at which the system achieves the final objective, typically indicated by the final task feedback. This includes not only the driver's final input action but also the processing and decision-making time of the HMI system after receiving the final command. For example, the endpoint of a task to play music is when the music starts playing; the endpoint of entering a destination and starting navigation is the beginning of route guidance; and the endpoint of opening the sunroof is when the sunroof starts to move. For some interaction tasks, the final feedback is at the hardware end of the vehicle's electrical system, and the time point at which the change in status occurs is difficult to perceive. Hence, the change in the system interface display can also be used to represent the change in the hardware status. For example, in a task to adjust the climate control temperature, the endpoint can be chosen as the change in temperature displayed on the screen, as the climate control system might not immediately change the fan speed and temperature upon receiving this command.

Response Time

Certain interaction tasks are not initiated by user input but are automatically triggered by factors such as the vehicle's location, status, and surrounding environment. For such tasks, the response time is used to determine the system's speed in recognizing triggering conditions, processing speed, and the timing of the interaction system output. The total time or total displacement along a particular direction must be calculated from the time point at which changes occur in the vehicle's position, status, surrounding environment, or other variables, to the time point at which the system shows the final task initiation feedback. For instance, if a vehicle does not follow the planned route to go straight at an intersection but instead makes a turn, how long does it take the system to detect the route deviation, re-plan the route, and resume guidance according to the direction after the turn? The response time of advanced driver-assistance functions is typically not within the scope of HMI evaluation but instead falls under the evaluation of the advanced driver-assistance system.

7.2.2 Operation Complexity

Operation complexity refers to the operational load of the driver when using the automotive HMI system. It can be comprehensively evaluated based on the number of operation steps, operational displacement, and fixation count.

Operation Steps

Operation steps refer to the total number of steps required by the driver to perform a specific function through a certain interaction modality. The measurement of operation steps typically begins with the central information display and instrument cluster display remaining on the system's home page and the voice control remaining inactivated. For other navigation tasks during the navigation guidance process, such as checking the route overview, turning off voice prompts, or adding waypoints, the initial state is selected as the navigation page with ongoing navigation guidance.

The calculation of operation steps varies for different interaction modalities. For some more ambiguous operations, it is necessary to manually establish a unified criterion to standardize the testing and evaluation process. Specific calculation methods can be referred to as follows:

- Touchscreen: Each finger tap, swipe, or hold is counted as one step. If typing or handwriting on the screen is not the main evaluation object, a continuous text input can be uniformly counted as one step. Otherwise, the large number of taps needed for inputting the text will occupy a significant proportion of the task steps, and the test results may downplay the issues related to excessive operation steps caused by poor interaction design.
- Central console buttons: Each press or turn is counted as one step. For knob-type central console buttons, if the rotation of the knob adjusts continuous numerical values (such as adjusting the climate control temperature), a single rotation is counted as one step. If rotating the knob is for selecting between different function modes (such as driving mode selection), each notch rotated is counted as one step.
- Steering wheel buttons: Each press or swipe is counted as one step.
- Voice control: Each continuous, uninterrupted sentence is counted as one step. Activating the voice assistant counts as the first step.

The methods described above for counting the operation steps are easy to implement and yield standardized results. Hence, it is an important index for examining the operational loads of physical movement and verbal dialogue. However, the operational load of each operation step is not fully equivalent. For instance, if we compare the action of tapping an icon on the screen with the action of tapping and dragging the same icon to a specific location, the latter is a continuation of the former and hence has a higher operational load. Similarly, the action of tapping a specific icon on the screen and dragging it to a particular location imparts a heavier operational load compared with simply swiping the screen to turn a page, as the former action requires precise positioning while the latter only needs approximate positioning. In terms of voice commands, saying "navigate to the P5 parking area, Helsinki Airport" represents a higher operational load than simply stating "start navigation", as the former involves more content and a mixture of words and numbers. To achieve precise measurements of operational load, different operation actions should be categorized under the various interaction modalities to conduct more in-depth research and experiments, and corresponding coefficients must be assigned for result adjustment. Although such an approach is more rigorous, it could potentially affect the convenience of implementing testing and evaluation and the standardization of the results.

Operational Displacement

Operational displacement refers to the total distance that a user's finger moves on the screen while performing a function via the touchscreen. The measurement begins from the moment the finger enters the screen area and ends with the final tap. For touchscreen interactions, in addition to the operation steps, operational displacement is a critical index for the physical workload imposed on the user. This is because the distance between each icon tap on the screen can vary significantly, with some being closer together and others being further apart. Longer distances necessitate not only a finger tap but also substantial arm movement. Operational displacement is typically not considered for physical buttons and voice control, as physical buttons usually have a concentrated distribution for a specific function, whereas voice control do not involve any hand movements.

The measurement of operational displacement involves sequentially connecting the starting point with the touchpoint of each icon until the final point, followed by calculating the total length. For icons that are not large, the geometric center of the icon can be chosen as the touchpoint. However, for icons with substantially larger touch areas, selecting the geometric center contradicts common user behavior. There are two common types of icons with large touch areas: the first includes buttons or text input boxes that are very wide horizontally, where users tend to tap the side closer to them rather than the center of the area, as shown in Fig. 7.1. The second includes the non-icon areas of the screen. For instance, in some vehicles, the destination search box of the navigation system is not always on the map page and appears only after a non-icon area of the map is tapped. In such cases, for that particular step, the non-icon area on the map becomes a very large "icon".

To more accurately reflect the user's real touch location and standardize the evaluation test, we propose a touchpoint selection method for large touch areas, as shown in Fig. 7.2. We denoted the touch area as S and the touchpoint to be selected as A.

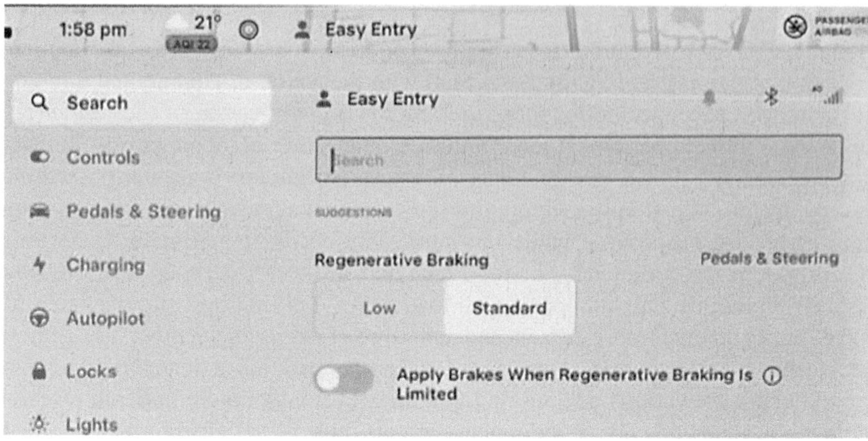

Fig. 7.1 Search box in the Tesla Model Y (picture taken in 2023)

When the horizontal width of area *S* is less than 10 cm, the horizontal coordinate of point *A* is at the midpoint of the area's width. When the width exceeds 10 cm, the coordinate of point *A* is 5 cm from the side closer to the user, i.e. 5 cm from the left for a left-hand-drive vehicle. When the vertical height of area *S* is less than 10 cm, the vertical coordinate of point *A* is at the midpoint of the area's height. When the height exceeds 10 cm, the vertical coordinate for the touchpoint center of the icon in the next step (referred to as point *B*) needs to be considered. If point *B* is within the top 5 cm of area *S* (the blue shaded area in the right panel of Fig. 7.2), the vertical coordinate of point *A* is 5 cm below the upper edge of area *S* (the blue point in the right panel of Fig. 7.2). If point *B* is within the bottom 5 cm of area *S* (the green shaded area in the right panel of Fig. 7.2), the vertical coordinate of point *A* is 5 cm above the lower edge of area *S* (the green point in the right panel of Fig. 7.2). If point *B* lies between the top and bottom 5 cm of area *S* (the red shaded area in the right panel of Fig. 7.2) or if there are no more steps, the vertical coordinate of point *A* is at the midpoint of the area's height (the red point in the right panel of Fig. 7.2).

The selection of the starting point for operational displacement distance is also critical. If the first step is considered as the starting point, it would be impossible to determine whether the overall location of the interaction area is far away from the user. This is especially the case for tasks with only one step, as the operational displacement would be zero, which would make comparisons impossible. Theoretically, the optimal starting point would be at the three o'clock position on the right side of the steering wheel, as the driver's right hand is usually at this position before reaching for the screen. However, this starting point presents practical difficulties in measurement. First, as the position of the steering wheel can be adjusted forward and backward or upward and downward, identifying an absolute standard position can be challenging. Second, as the steering wheel and central touchscreen are not on the same plane, the spatial distance cannot be easily measured. A simplified method is to define the starting point as the point on the user-side edge of the central

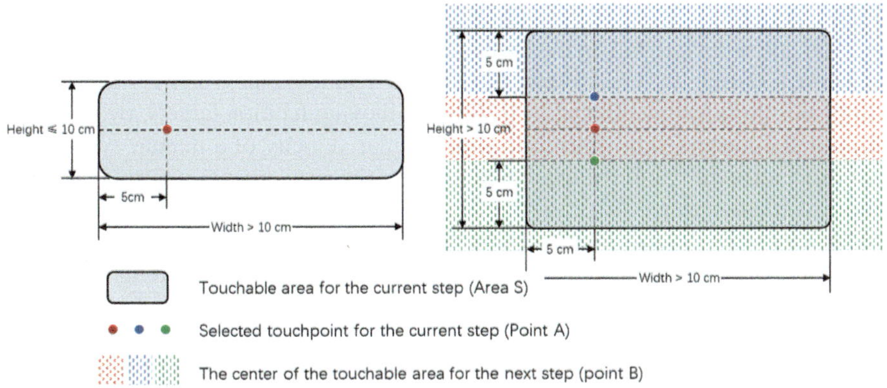

Fig. 7.2 Touchpoint selection method for icons with large touch areas when calculating the on-screen distance

touchscreen display (excluding the frame) at the same height as the icon for the first step. For left-hand-drive vehicles, this would be the left edge, as shown in Fig. 7.3. This method ignores the difference in the distance from the left side of the central information display to the right end of the steering wheel, potentially conferring an additional advantage to early market models with very small central information displays. However, for mainstream intelligent vehicles currently on the market, the variation in this distance is minimal.

Fixation Count

The fixation count refers to the total number of fixations on the interaction interface area (central information display, instrument cluster, or button area) when a certain function is performed via a specific interaction modality. As defined by the ISO-15007 international standard, fixations refer to brief pauses in eye movement within

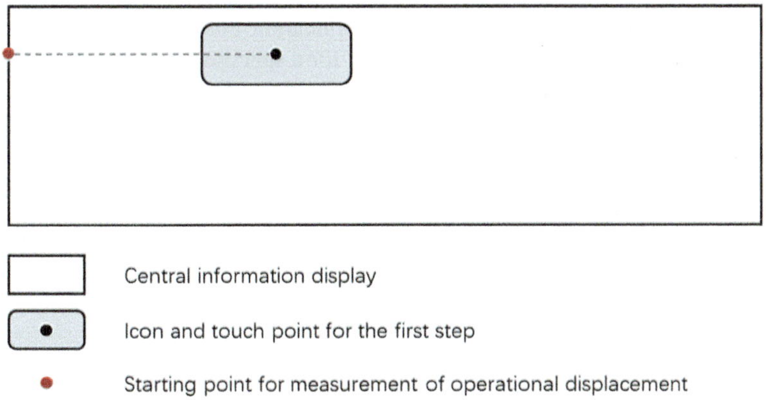

Fig. 7.3 Starting point selection for the measurement of operational displacement

a given period [1]. These pauses allow the eye to fixate on a particular point within the AOI, allowing it to fall on the fovea centralis of the retina. A typical fixation lasts between 100 and 2000 ms. When filtering for fixations, the Tobii I-VT algorithm eliminates those below 60 ms [2, 3]. For non-moving fixation targets, two fixations are usually connected by a saccade. The angular velocity of a human eye saccade can reach up to 500°/s. If the angular velocity of the gaze exceeds a certain threshold during a fixation (e.g., 30°/s as stipulated in the Tobii I-VT algorithm [2]), it will be split into two distinct fixations. The fixation count not only varies according to the fixation algorithm used but is also affected by the precision and reliability of the acquisition equipment. Typically, the results are comparable only when using the same eye-tracking equipment and the same fixation algorithm.

The fixation count primarily assesses the ease of finding an icon or a button. During a given step in executing an interaction task, it is often necessary to first locate the icon or the button to be tapped or pressed using the eyes, which is then followed by extending the finger in the direction of the eye's positioning. Generally, the eyes require multiple fixations to find the target location. For example, when using the touchscreen to perform the task of switching to the next music track, the driver's gaze moves away from the road to the central information display, which first produces two fixations in the navigation widget area before moving to the music widget area, as shown in Fig. 7.4. Within the music widget, a fixation first occurs on the switch music source icon in the lower-left corner before locating the next track icon in the lower-right corner. The easier it is to find the target, the fewer the fixations required for each operation step. Conversely, targets that are difficult to find require numerous fixations for each step. For some central console and steering wheel buttons with distinctive positions and tactile sensations, "blind operation" may be possible, that is, operation with no fixations.

To a certain extent, the number of fixations and operation steps are mutually restrictive. For example, as shown in Fig. 7.5, 12 icons are present in the middle of the left interface, which represent the entry points to 12 different applications, one of which is the Amap application (in the red rectangle). Three larger icons are present in the middle of the right interface, one of which is also Amap (in the red rectangle). Because the left interface provides more application entry points, the driver may avoid swiping to the next page when looking for a specific application to reduce the number of operation steps. However, the icons on the left are relatively small and numerous, making them difficult to locate. During the interaction step of tapping the Amap icon, the driver's fixation count on the left interface is higher than that on the right interface, and the time consumed is also longer. Thus, combining the fixation count with the number of operation steps to evaluate operation complexity can prevent design practices that arbitrarily increase the page content to reduce the number of operation steps.

Fig. 7.4 Fixation distribution when using the touchscreen to switch to the next music track in the Li Auto One

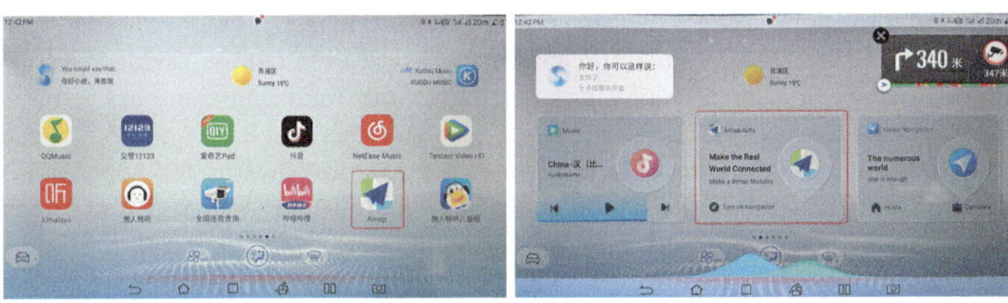

Fig. 7.5 Different "Amap" sizes (in red rectangles) on different pages in the central information display of BYD Destroyer 05 (2023 Model)

7.3 Summary of the Evaluation Results and Design Suggestions

7.3.1 Comparison Among Typical Modalities

Each operation step within an interaction task exhibits different time and fixation count distributions across various interaction modalities. Understanding the operation time structure and fixation distribution of various interaction modalities can provide valuable insights into the design optimization of HMI systems.

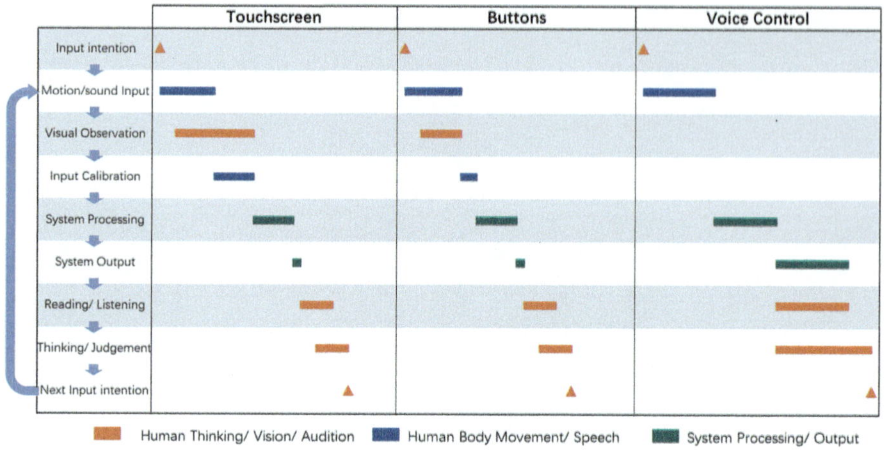

Fig. 7.6 Schematic of a typical task time distribution across various interaction modalities

Figure 7.6 shows the time distribution for a driver operating interaction tasks using various interaction modalities. The process from top to bottom represents one operation step within the task, which then continues to cycle; if the task has several steps, it will cycle several times as shown in the diagram. The length of the colored rectangular block represents the time consumed for that step.

During touchscreen interactions, the driver's operation can be divided into the following steps: first, the driver forms an input intention for this interaction step based on the interaction task and determines the icon shape or text content to be tapped. Second, the hand leaves the steering wheel and extends towards the approximate direction of the target location on the screen. Third, just as the hand begins to extend, the eyes will look at the screen to search for the precise target location to guide the direction of touch. Fourth, the movement of the finger is calibrated based on the visual feedback from the screen, and the target location is tapped, which marks the completion of the physical input for this interaction step. Fifth, the HMI system processes the information after receiving the driver's input. Sixth, the HMI system outputs information on the screen or through speakers, such as entering the next layer of menu on the screen. Seventh, the driver reads or listens to the HMI system output. Eighth, the driver thinks and judges the output information seen or heard. Ninth, the driver forms the input intention for the next interaction step; subsequently, the next cycle begins. Touchscreen interaction differs from physical button interaction in two ways: (i) owing to the large amount of information on the screen that constantly changes on different pages, finding a specific icon is more challenging; (ii) as the screen surface is flat and smooth, the driver cannot confirm whether the tapping location is correct by touch alone, and accurate touching of the target location is only possible with continuous visual support. These factors contribute to longer gaze diversions and more fixations.

When using physical buttons, the process is approximately the same as touchscreen interactions. However, as the buttons are usually large and have distinct tactile feedback, the button's shape and edges can help the users identify whether the button-press position is correct. Therefore, users can quickly adjust the finger position based on touch alone without continuous visual support. Compared with touchscreens, the button interaction process significantly reduces the time consumed while also contributing to shorter gaze diversions and fewer fixations. For poorly designed buttons, such as those with small areas or unclear edge tactile feedback, constant visual support may also be needed. For well-designed steering wheel buttons, "blind operation" can often be achieved, i.e., without the need for visual observation, as only finger movement is needed instead of arm movement.

Voice control differs from other interaction modalities. If we do not consider the button trigger for voice conversations (most new models can be voice-activated), drivers can observe the road before speaking without looking at the screen. Moreover, under correct voice recognition, the speech content is not necessary to be calibrated by users. However, the time consumed by speaking is relatively long; even a simple four-word command usually takes approximately 2 s. During the system processing phase, the system needs to first comprehend the semantics and then input to system for analysis and judgement, which is also time-consuming. During the system output process, the system has to read out the content, which requires even more time. However, listening to the system output or reading the content displayed on the screen can be performed synchronously with the system output process and, hence, does not require additional time. Finally, the thinking/ judgement process for the intention of the next interaction step are similar to those of the touchscreen and button interactions. The total time consumed for each interaction step in voice interaction is relatively long. This is primarily because voice information is one-dimensional linear, which has a relatively low input and output information efficiency as well as more complex system processing. For voice control systems that support interrupted conversations, if users can predict the output content before the system has finished reading, they can interrupt the reading and directly say the next sentence, that is, enter the next interaction step. This can compress the system output time and overlap the thinking/ judgment time with the system output, thereby reducing the total time required for this interaction step. Although each voice interaction step requires a longer time, it does not imply that voice interaction is necessarily inefficient. For more complex interaction tasks, voice control can often achieve the objective through only two or three steps, whereas touchscreens and buttons require more steps to accomplish this.

7.3.2 Relationship Between Efficiency and Safety

Efficiency and safety are two distinct and independent indexes. However, in actual test results, the two often exhibit correlation. A longer task time usually leads to a longer total gaze diversion time because most tasks require intermittent screen observation throughout the process. More operation steps usually lead to larger lane

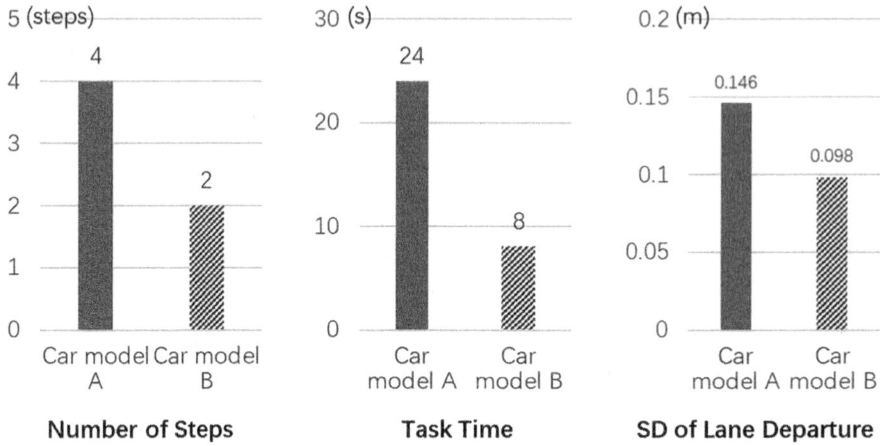

Fig. 7.7 Data comparison between efficiency and safety when executing the "navigate home" task via voice control

departures, because each physical action will cause instability in steering control, and the lane departures caused by these instabilities will accumulate as the number of operation steps increases. For example, as shown in Fig. 7.7, when executing the "navigate home" task with voice control, car model A has more operation steps and requires a longer task time compared with car model B, which results in a larger standard deviation of lane departure for the former.

Efficiency and safety are not always positively correlated. For instance, let us consider two steering wheels: one with fewer buttons that can be operated blindly and another with more buttons that is harder to operate blindly. When completing a certain interaction task, the former steering wheel may need two steps, whereas the latter only needs one. In this condition, the former may be less efficient but safer. The correlation between efficiency and safety can be even weaker under different interaction modalities. For example, voice control might take longer than touchscreen control for some tasks, but gaze diversion requires less time during voice interaction, and there are no manual distractions. Hence, voice control might be safer than touchscreen control.

7.3.3 Suggestions for Interaction Modality Selection

By conducting a comprehensive analysis that combines the efficiency and safety evaluation indexes, we can provide suggestions for interaction modality selection for different categories of interaction tasks, which can serve as a reference for automotive HMI design.

For simple tasks typically involving one or two steps using touchscreens or buttons (e.g., adjusting the music volume and answering phone calls), the following are recommended:

- Steering wheel buttons as the optimal interaction modality. Design optimizations, especially those that reduce the number of operation steps, can achieve better efficiency and safety.
- Central console buttons should also be available. On one hand, driver distraction caused by central console buttons is only slightly higher than that caused by steering wheel buttons. On the other hand, central console buttons have a more extensive range of functions and can therefore be applied as a redundant and supplementary to steering wheel buttons.
- Touchscreens need not be involved in these tasks. They cause significant driver distraction and longer task times. If physical buttons are inadequate for these tasks and the use of touchscreens is necessary, the design should be optimized as much as possible to ensure that the icon has a fixed position with a reasonable size, and can be located quickly and easily.
- Voice control also need not be involved in these tasks. For these simple tasks, its advantages of fewer operation steps and shorter gaze diversion time are not fully performed; instead, it highlights the strong cognitive distraction of voice control. However, for the sake of functional completeness of the voice control system, the inclusion of these simple tasks can also be considered, especially because voice control does not have a visible interface, and adding certain functions will not encroach on the layout space of other functions.

For intermediate tasks, typically involving 2–4 steps using touchscreens or buttons (e.g., calling a designated contact and switching audio sources), the following are recommended:

- Voice control is the optimal interaction modality. It requires significantly less visual attention compared with other interaction modalities and can usually accomplish the task in a single step.
- Although touchscreens and central console buttons can increase driver distraction, at least one of them should be retained for controlling intermediate tasks. This is because they have better learnability and offer a more extensive range of functions.
- Steering wheel buttons need not be involved in intermediate tasks. Owing to their limited number, they might require more steps if used to complete such tasks.

For complex open-input tasks (e.g., entering a destination and starting navigation, or inputting a phone number and making a call), the following are recommended:

- Voice control is the optimal interaction modality. It requires less visual attention, can reduce the number of interaction steps, and does not necessitate the use of complex keyboards or handwriting to input open-ended text content.
- Touchscreens should also include these task types. Although they impose higher visual demands, longer task times, and more operation steps, their learnability and functionality richness are superior to those of voice control.

- Central console and steering wheel buttons need not include these task types. The process of inputting open text content is extremely complex with these modalities, leading to poor performance in terms of both efficiency and safety. Thus, it is unnecessary to allocate limited and valuable physical button resources for these tasks.

References

1. Road vehicles — Measurement and analysis of driver visual behaviour with respect to transport information and control systems: ISO 15007: 2020 [S].
2. Tobii Technology. The Tobii I-VT Fixation Filter Algorithm description [R]. 2012.
3. Jarkko S, Kai P, Jaana S, et al. Inferring Relevance from Eye Movements: Feature Extraction [Z]. Helsinki University of Technology, 2005.

Chapter 8
Cognition

8.1 Development

Cognition is an index originating from cognitive psychology. Within the context of automotive HMI systems, it refers to the system's facilitation of the users' ability to perceive, comprehend, remember, and apply information correctly and efficiently during use.

Cognitive psychology is a research domain in psychology that investigates how individuals attend to and select information, cognize and store information, make decisions based on the information obtained, and generate external behaviors. Its purpose is to explain and clarify how information processing occurs during cognitive activities, how external information is stored in the mind, what information is utilized when solving problems, and what type of thinking strategies are adopted. Therefore, cognitive psychology can also be referred to as information-processing psychology [1].

Cognitive psychologists have conducted extensive experiments to investigate human cognition, primarily focusing on perception, attention, knowledge, language, memory, and thinking, encompassing both the physiological (such as changes in receptor potential) and behavioral (such as recognition-by-components theory) levels. The conclusions drawn are scientific and universal, providing a valuable reference for other studies related to human factors [2, 3].

Perception is the experience generated by the stimulation of sensory organs. In daily life, over 80% of human perception originates from vision. In cognitive psychology experiments concerning vision, target stimuli with different physical properties such as different sizes, colors, and shapes are often used. These stimuli can be two-dimensional items, such as texts and images, or three-dimensional entities. Researchers analyze the differences in the visual recognition of target stimuli with different properties to derive the general principles of human visual recognition. Although other types of perception beyond vision are also within the scope of research, different experimental methods are employed.

J. Ma and Z. Gong, *Automotive Human-Machine Interaction (HMI) Evaluation Method*, Research on Automotive Intelligent Cockpit,
https://doi.org/10.1007/978-981-99-9951-4_8

Attention refers to the ability to focus on specific stimuli or locations. For example, when one is reading a passage, one's attention is allocated to the text. Attention is directed and concentrated. At a given moment, human mental activity selects a particular target and concentrates on it, ignoring others. People can only pay attention to one thing at a time. Cognitive psychologists have continuously summarized and refined attention allocation models via experiments, allowing them to draw conclusions regarding human attention allocation mechanisms, many of which can be applied to the field of automotive interaction design.

Knowledge refers to various mental representations used in cognitive processes, including memory, reasoning, language use, and comprehension. Current research on knowledge is primarily centered on categorization. Correct categorization can facilitate people's comprehension of things. Herbert Simon, an American psychologist, believes that human cognition of things is the cognition of relationships. In cognitive psychology, Gestalt's theory is widely applied in interaction design.

Language is composed of a system of sounds and symbols used to express feelings, ideas, thoughts, and experiences. In cognitive psychology, research on semantics, syntax, conversation, and language culture can provide theoretical support for studies related to voice interaction in HMI.

Memory refers to the cognitive process of retaining, retrieving, extracting, and using information in the absence of the initial information (e.g., stimuli, images, events, ideas, or skills). Cognitive psychology examines short- and long-term memory, which have yielded relatively quantitative findings on the models, capacity, content, and duration of human memory, thereby providing practical guidance for interactive content design.

Thinking is the cognitive process of problem-solving, reasoning, and decision-making based on external information. Cognitive psychology explores the patterns and processes of human thinking through specific experimental methods to explain the influencing factors behind human behavior, thereby providing a scientific reference for researchers in the field of human factors.

Several components of human perception and behavior control are processed in parallel. However, processes such as memory, thinking, and attention at the cortical level of the brain are mostly processed in series. Humans can only do one thing in 0.25 s, which is serial processing, akin to single-threading in computers, meaning that any time elongation in any of these processes will result in an extended total duration of cognitive activities.

Research on HMI is inextricably linked with cognitive psychology. Based on the model of human information processing and the basic principles of computer science, we can derive the information processing model for an HMI system, as shown in Fig. 8.1. The human information processing model postulates that once humans receive stimulus information, it undergoes processing through the perception, cognition, and response systems, eventually resulting in actions. The fundamental principle of computer operation involves the input device receiving raw data, which then undergoes computation and storage to be transformed into an output format that can be received by humans. Throughout the entire HMI process, users obtain information from output devices such as screens and speakers. After being

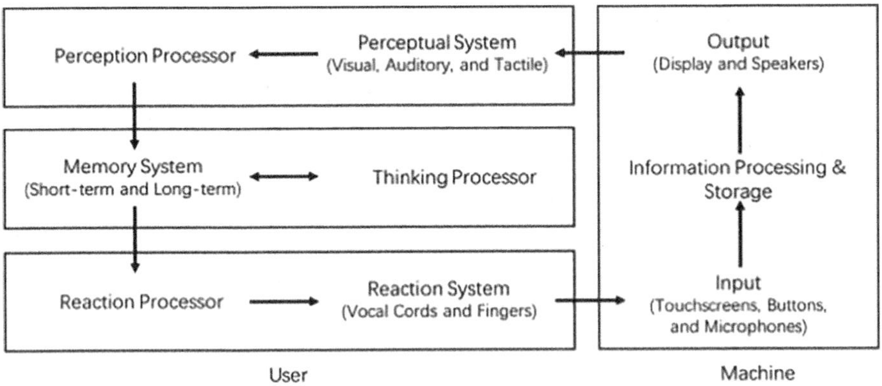

Fig. 8.1 Information processing model of an HMI system

stored and transformed, this information can guide users in making decisions and responses, such as pressing a specific button or answering a particular question.

It is vital to study and evaluate the automotive HMI in accordance with the theories of cognitive psychology. Even a simple one-step task, such as switching to the next track using a touchscreen, involves several aspects of cognitive psychology. First, the user needs to visually observe various information and function icons on the screen. Next, based on the properties of the icons, such as shape and color, the user needs to identify which icon represents the "next track" function and extend their finger to tap on the icon. If this task is performed while driving, the user needs to observe the road conditions for safety before looking at the screen and quickly return their gaze to the road after tapping to continue driving. Within a few seconds, the user's cognition undergoes several processes: perception (looking at the road and screen), attention (allocating attention to driving and interaction tasks), knowledge (recognizing a safe environment and correct icons), memory (remembering the icon position), and thinking (determining when is the right time to tap the icon to complete the task). It is only by breaking down the tasks according to cognitive principles that we can precisely identify the fundamental issues in automotive HMI design and perform necessary optimizations and improvements.

Some research results in cognitive psychology are quantified through subjective rating scales. An example of this is the standard System Usability Scale (SUS), which consists of 10 items, as shown in Table 8.1. Except for the first item, the other items are directly or indirectly associated with the principles of cognitive psychology. The advantage of using standardized rating scales is that they allow for cross-temporal and cross-disciplinary comparisons. However, these scales have limitations when applied to the cognitive evaluation of automotive HMI systems. First, these questionnaire scales are subjective. The scoring for each item, such as when to assign 3 or 4 points, is not specifically defined but depends entirely on the respondent's interpretation. Second, these questionnaire scales target all fields, including smartphones, computer software, washing machines, and refrigerators. Consequently, the

Table 8.1 Standard SUS scale

		Strongly disagree			Strongly agree	
		1	2	3	4	5
1	I think I would like to use this system					
2	I found the system unnecessarily complex					
3	I thought the system was easy to use					
4	I think I would need the support of technical personnel to be able to use this system					
5	I found that the various functions in this system were well integrated					
6	I thought there was too much inconsistency in this system					
7	I would imagine that most people would learn to use this system very quickly					
8	I found the system very cumbersome to use					
9	I felt very confident when using this system					
10	I needed to learn a lot of things before I could get going with this system					

items tend to be fairly general and lack specificity when applied to complex systems such as automotive HMI.

An objective, quantitative evaluation system must be established for the cognitive evaluation of automotive HMI, complemented by relatively subjective methods for indexes that are difficult to objectify completely. Compared with other daily human activities, the information and possible operations within automotive HMI systems are finite, closed, and exhaustible. Therefore, when quantifying from a cognitive psychology perspective, fewer influencing factors need to be considered. Furthermore, the process of operating an automotive HMI system is directly related to several quantitative findings in cognitive psychology, such as the short-term memory capacity of the human brain and the time it takes to process information. This provides a clear basis for quantitative evaluations. For example, if the amount of information on the central information display exceeds the short-term memory capacity, it would be considered suboptimal in this dimension.

The cognitive evaluation of automotive HMI is not only important but its standards are also more stringent than those in other fields. When a driver operates the automotive HMI system while driving, attention is divided between the primary task of driving and secondary interaction tasks. Tasks with a high cognitive load not only have a low execution efficiency but can also affect driving safety. Therefore, it is necessary to accurately quantify the cognitive evaluation standards for automotive

HMI systems, define the design boundaries, and control the cognitive load of the user within a safe range.

8.2 Evaluation Indexes

The cognitive evaluation of automotive HMI can be subdivided into second-level evaluation indexes, including logical structure, element visibility, element understandability, element memorability, and system feedback.

8.2.1 Logical Structure

Logical structure refers to the clarity and ease of use of the logic among different hierarchical levels and among different elements of the automotive HMI interface. If an automotive HMI has an excellent logical structure, it will be easier for users to learn, and users will be more willing to use it. The logical structure is strongly associated with knowledge, memory, and thinking in cognitive psychology. Logical structure is primarily divided into the hierarchical structure of the overall information and the clustering and subordination relationship of specific elements.

Hierarchical Structure

The hierarchical structure of automotive HMI systems primarily targets the various types of information within the central information display. This is because the central information display in contemporary vehicles typically brings together the majority of HMI functions, which are divided into levels according to a certain logic. For function buttons that can interact independently without relying on the screen (excluding directional buttons that operate in conjunction with the screen), a hierarchical structure is not present because each key has only a single function. Voice interaction can also reach a specific function directly via a command and, hence, does not have a significant hierarchical structure. The instrument cluster display and the HUD also have a hierarchical structure, but they are generally significantly simpler than the central information display.

Evaluating the hierarchical structure of automotive HMI systems is important for two main reasons: first, automotive HMI systems have a wide array of functions, with the central information displays of several vehicles managing more than 1000 functions. Second, the hierarchical structures of automotive HMI systems can vary significantly across vehicles, each exhibiting varying strengths and weaknesses.

The interaction hierarchy of the central information display can be divided into three layers: the root directory, application, and shortcut layers, as shown in Fig. 8.2. The root directory layer serves as the entry point to all features and applications. It could be a group of icons representing different function modules or an arrangement matrix consisting of dozens of function applications, as shown in Fig. 8.3. Notably, the

Fig. 8.2 Three hierarchy layers of the Apple CarPlay interaction structure

root directory is not necessarily the homepage of the central information display. The application layer consists of every specific application under the root directory layer, including maps, music, telephone, climate control, and movie ticket reservation, each of which is a relatively independent application. Each application might have several internal layers, but they are not the focus of the overall hierarchical structure of HMI systems. If all functions must be operated or read from the root directory layer to the application layer, the steps could be complex, potentially leading to increased driver distraction. Thus, it is possible to extract some frequently used operations and display functions to create a shortcut layer. This layer can include dynamic widgets on the homepage or control interfaces, which can be activated by imprecise taps or swipes, such as the quick activation of drop-down or right-slide menus, as shown in Figs. 8.4 and 8.5.

Owing to the significant differences in the hierarchical structure across different car models, using a single objective evaluation method to cover every aspect of the related design is challenging. However, by surveying mainstream models on the current market, we summarized some issues that should be avoided in hierarchical structure design:

(a) Absence of a clear root directory layer: the root directory serves as the entry point for users to access various features as well as the starting point for users to understand the hierarchical logic. If no clear root directory exists or if several features cannot be accessed through the root directory, users may find it difficult to determine where to start looking for certain features.

(b) Lack of a clear shortcut layer: the shortcut layer can significantly improve the operational efficiency of relevant tasks and reduce driver distraction. Without a shortcut layer, most features would only be located in deeper hierarchies, making them difficult to find and complex to operate.

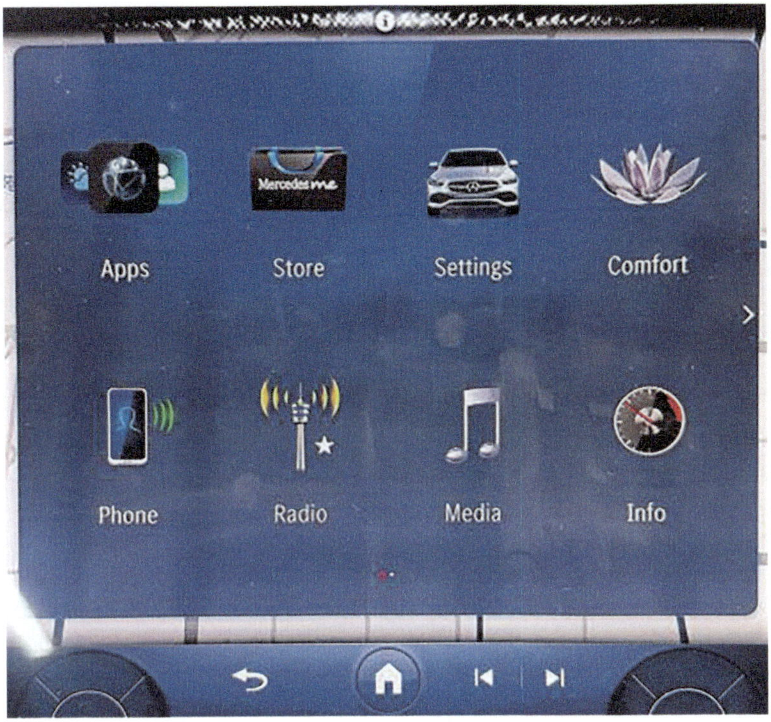

Fig. 8.3 Root directory of Mercedes-Benz C-Class (2023)

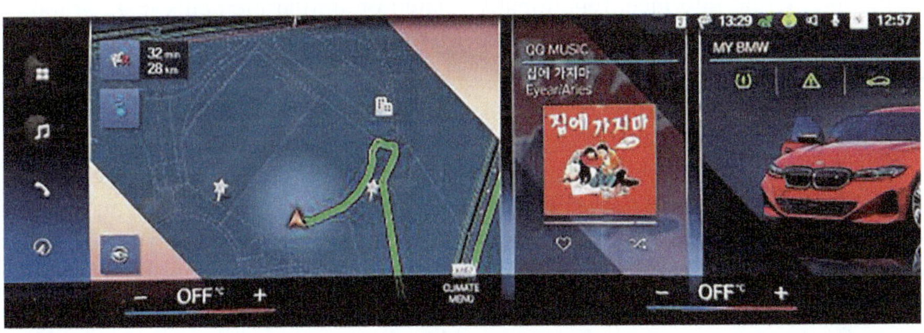

Fig. 8.4 Real-time navigation and music information displayed on the home page widgets of BMW i3 (2022)

(c) Inability of the shortcut or root directory layer to display the real-time status of certain applications (e.g., navigation and music). It should display the name of the current song and album cover in the music application, and the route taken by the vehicle and guidance information in the dynamic widgets on the

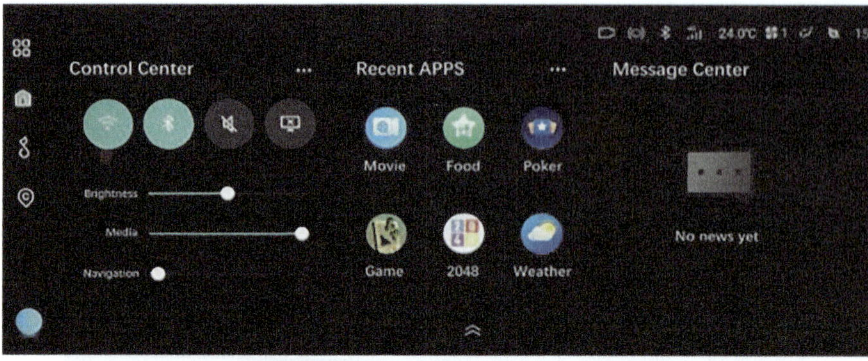

Fig. 8.5 Drop-down settings menu of Changan UNI-T (2021), translated from Chinese language

home page. This allows users to check important information from different applications at all times without having to enter each specific application.

(d) Presence of multiple nested systems with inconsistent operation logic: some automotive HMI systems are developed by different teams and then nested together such that each system may have a significantly different operation logic. For example, one system's directory may be a grid, whereas another's is a vertical list. Thus, users will need to switch between different operation logic, which adds unnecessary cognitive load.

Element Clustering and Subordination

In the interaction interface, functionally related or similar elements (e.g., graphics, icons, or buttons) should be grouped together. They can appear on the same page or module on the screen or within a relatively independent area on the central console or steering wheel, which facilitates the user's comprehension and memory for the layout of similar functions and enables quick searching. For example, climate control functions should be clustered together, as should the lighting-related functions. Similarly, lighting functions should not appear within the climate control module, and climate control functions should not appear outside of their designated module. However, element clustering does not imply that one function type can only be grouped in one area; they can also be clustered in modules on 2–3 different layers based on the usage frequency or scenario. Let us consider climate control as an example. Here, commonly used functions such as temperature and airflow adjustment can be fixed in the lower area of the central information display or made to appear in a drop-down menu, whereas complete climate settings can appear in a separate interface.

In an interaction interface with multiple layers, the elements appearing in the next layer should be definitively and indisputably subordinate to the previous layer. Only a logical structure with correct subordination relationships can allow users to confidently decide which icon to click on the previous level to identify the desired final function in the next layer without misunderstanding or causing errors in interaction paths. For example, music streaming should be subordinate to the music module, and the vehicle's driving assistance feature settings should be subordinate to the vehicle settings module. If all climate control functions are placed in the vehicle settings module, this will go beyond the experience of general users, leaving them uncertain about where to search for climate control functions.

Element clustering and subordination are basic requirements for the logical structure of an automotive HMI system, and most car models have no issues in this regard. However, once such problems arise, they not only increase the cognitive load of specific functions but may also severely damage the user's overall impression regarding the HMI system.

8.2.2 *Element Visibility*

Element visibility refers to the degree to which icons and text in the interface can be clearly seen by the user when executing an interaction task. Element visibility is closely related to perception in cognitive psychology. Whether an element can be clearly seen largely depends on its size; however, it is also related to its color difference from the background.

Icon and Text Size

Icons and text need to reach a sufficient size to be clearly seen by the user so that observational errors can be avoided and excessive attention would not be drawn. The displayed size of icons and text cannot be expressed in font size because the same font size can be displayed differently on screens with different pixel densities. The most accurate measurement for the font size is the visual angle subtended by the icon or text at the human eye, measured in arcminutes (60' = 1°). The relationship between the visual angle and the actual symbol height is shown in Fig. 8.6 and is expressed as follows [4]:

$$H = 1000 \cdot D \cdot \tan\left(\frac{V}{60}\right)$$

In actual tests, measuring visual angles is usually a complicated process. As the distance from the human eye to the central information display is similar in different car models, usually between 0.75 and 0.85 m, this value in the equation mentioned above can be considered to be 0.8 m, and the reference value for height H can be directly obtained. A separate measurement for the visual angle is only needed for car models with exceptionally unique screen layouts.

In 2016, the U.S. National Highway Traffic Safety Administration (NHTSA) proposed that the optimal visual angle for text should be 20', corresponding to a height of approximately 4.7 mm, and the minimum visual angle for text should be 12', corresponding to a height of approximately 2.8 mm [5]. The optimal visual angle for icons should be 86', corresponding to a height of approximately 20 mm, and the minimum visual angle for icons should be 34', corresponding to a height

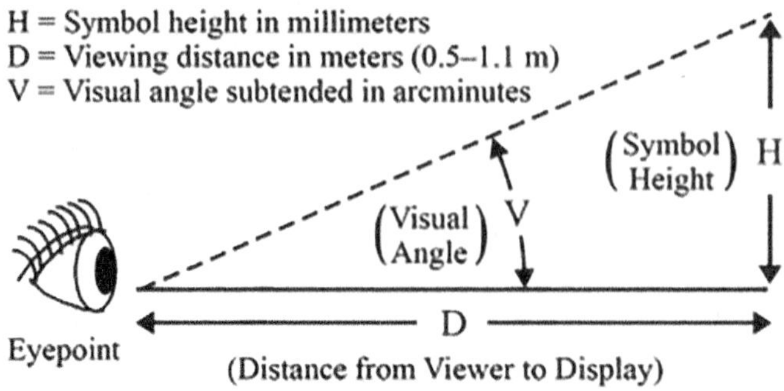

Fig. 8.6 Geometric relationship between the symbol height and the visual angle

of approximately 8 mm. The NHTSA design guidelines can be used as a reference; however, adjustments are needed for the current intelligent vehicle market in China. On one hand, the stroke density of Chinese characters is higher than that of English characters; therefore, the size at which Chinese characters can be clearly visualized should be larger than that of English characters. On the other hand, on high-resolution screens, icons are presented in various forms. Sometimes a simple icon may be similar to text in height and can be clearly distinguished; hence, the standards for certain forms of icons can be more lenient in the evaluation process.

Background Color Difference

In addition to the appropriate size, the icons and text on the central information display should also have a significant color difference from their background such that users can easily see them. This is especially true when the icon or text background is a map or a more complex image, which is more likely to result in a poorer color difference between the background and the icon or text. The difference between two colors can be represented by the numerical value of ΔE, wherein a difference greater than 100 is considered significant. However, in actual tests, accurately measuring the element color on the screen is relatively challenging. Third-party testers usually cannot get the original interface design files and, therefore, cannot directly read the color values of the elements. Moreover, the color displayed on the screen may differ from that in the design file. Hence, the ideal method is to use expensive dedicated equipment to directly identify the colors displayed on the screen. Ordinary colorimeters cannot measure luminous screens. If a camera is used to directly photograph the screen for analysis, the colors in the photographs often deviate significantly from the actual colors displayed on the screen.

Icon Color Difference

During the visual search for a specific target, individuals can often locate the target more quickly if they can first determine the target's color. In the rapid search for a specific target, color is frequently a more prominent characteristic than shape. For

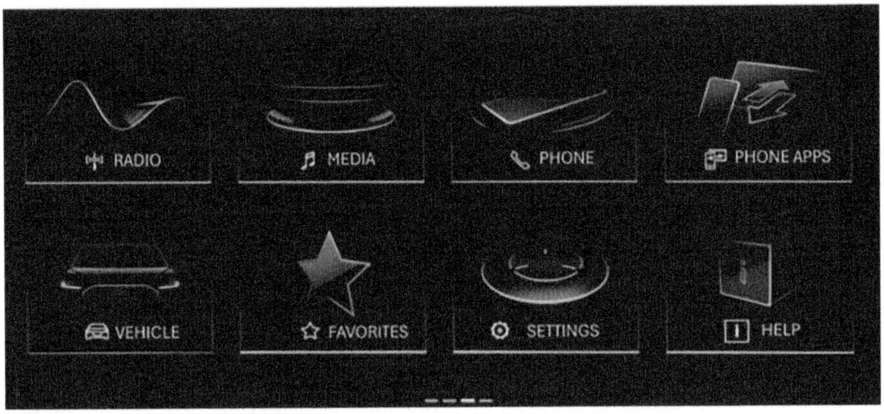

Fig. 8.7 Colored bottom edge of icons on the home page of the central information display in Audi A3 (2021), translated from Chinese language

instance, when looking for a particular banknote denomination in a wallet, we first distinguish the banknotes by color rather than the patterns and numbers on them.

The same principle applies in the context of the home page and root directory in a vehicle's central information display. The use of icons with different colors can considerably aid users in conducting rapid searches. Some vehicles employ a uniform color design for all icons, forcing users to rely on the icon shapes and adjacent explanatory text to make distinctions, which can reduce the efficiency of visual searches. However, the difference in icon colors does not necessitate the application of a particular color to the entire icon, as this could potentially lead to a chaotic design and hence compromise the esthetics. Embellishing certain details on icons using colors can serve as a guide for the visual search, for instance, the lower edge of an icon border, as shown in Fig. 8.7. Typically, the presence of color variations in the icons is sufficient to produce a significant color difference; hence, there is no need for specialized equipment for color measurement.

8.2.3 Element Understandability

Element understandability refers to the extent to which elements in the interface align with the user's common knowledge and are easily understood by the user when performing interaction tasks. It encompasses the ease of understanding icons and texts as well as the ease of understanding interaction states. Element understandability is closely associated with knowledge, language, and thinking in cognitive psychology.

Icon Meaning

In the interface of an automotive HMI system, icons should be designed in such a way that the users can easily associate them with their corresponding functions.

Designers should avoid creating overly abstract icons or completely deviating from conventional icon designs for the sake of being unique or different. For icons related to vehicle control and indication, reference can be made to the ISO-2575 international standard, as presented in Table 8.2 [6]. Other icons related to vehicle navigation, communication, and entertainment can be based on common designs found in consumer electronic products such as smartphones.

Certain conventional icon features are commonly used in smartphones and other consumer electronic devices. For example, in the case of an incoming call, the green icon represents answering the call, whereas the red icon represents rejecting it. On smartphones and computers, the phone, folder, and email icons are typically green, yellow, and blue, respectively, as shown in Fig. 8.8. Additionally, some popular applications have unique colors, which have become an important part of user recognition and branding. For instance, the icon of WhatsApp is green, and TikTok is black. If icons of different colors are used in the automotive HMI system interface, they should be consistent with the conventional color features of these icons. Otherwise, this may cause confusion or misunderstanding among users. If all icons on the interface are the same color, the design conventions of these icon colors need not be considered; however, such a design may lead to deficiencies in element visibility.

Textual Representation

In automotive HMI systems, the displayed text and announced audio content should be accurate and free of ambiguity, spelling errors, or polyphonic mispronunciations. Textual errors are considered relatively basic errors, and most mature car models in the current market do not have such issues. Imported car models require translation of the text in the HMI system when introduced to the Chinese market. Some translations are extremely direct and lack localization, which may lead to misunderstandings among Chinese users. Similarly, exported cars from Chinese brands sometime fail to translate wordings correctly. For example, vehicle's climate control system is usually translated as air conditioning system, which is direct corresponding to the word in Chinese language.

Information Visualization and Cross-Reality Operation

Information visualization utilizes methods such as graphics, images, and animations to help users understand and analyze information. When driving, users need to pay attention to several pieces of information. If all the information is presented in the form of numbers and text, users would need to first read and understand these numbers and text, then interpret what the information means, and finally decide on their next action. This entire process can impose a heavy cognitive load. By contrast, effective visualization methods can directly convey the meaning of the information to users, enabling them to make quick decisions and reduce the cognitive load. For example, the distance in the navigation prompt shown in Fig. 8.9 requires users to estimate how far 184 m is, whereas the blue countdown progress bar in the background intuitively conveys how far the ramp is. In Fig. 8.10, the visual effects of vehicle lights at the top will dynamically change according to the settings in the menu below, helping users understand the specific functions of different light modes. Information visualization

Table 8.2 Summary of the icons related to vehicle control and indication in the ISO-2575 international standard (excerpt)

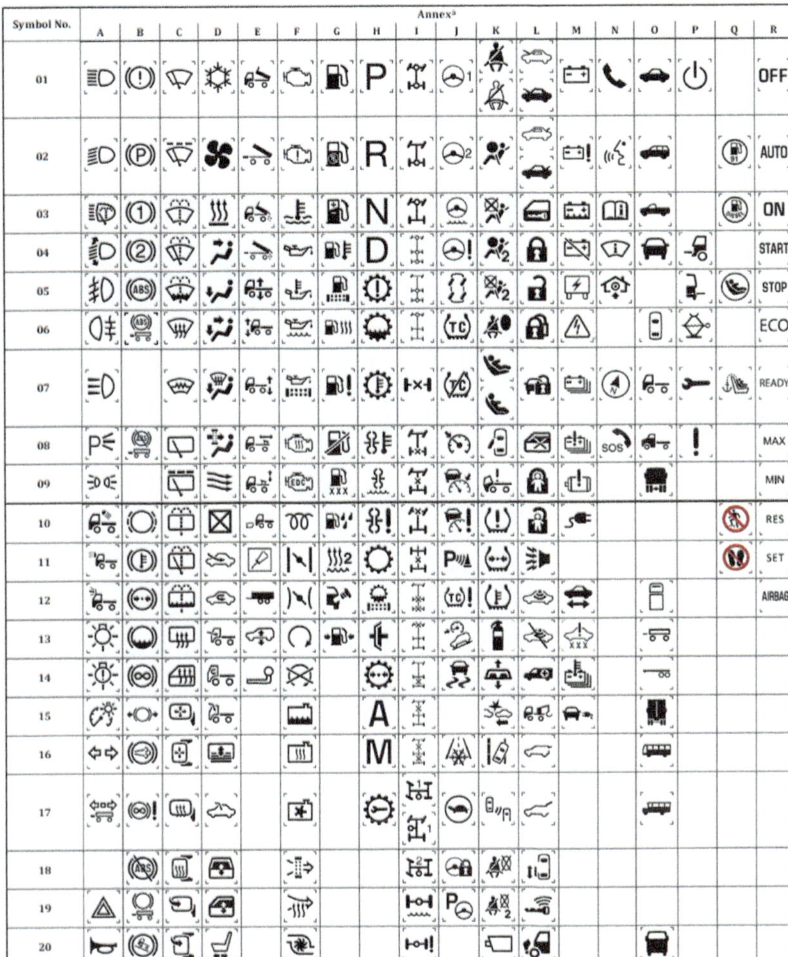

[a] A = Lighting and signalling devices
B = Braking systems
C = Visibility
D = Cab environment and comfort
E = Maintenance and load functions
F = Engine
G = Fuel system
H = Transmission
I = Power drive
J = Vehicle handling and cruise control
K = Active and passive safety systems
L = Security
M = Electric functions in general and electric road vehicles
N = Information and communication
O = Generic vehicle shapes
P = Miscellaneous
Q = Special symbols
R = Special signs
S = Additional informative symbols

Fig. 8.8 Some icon designs on Xiaomi MIUI 12 (2020)

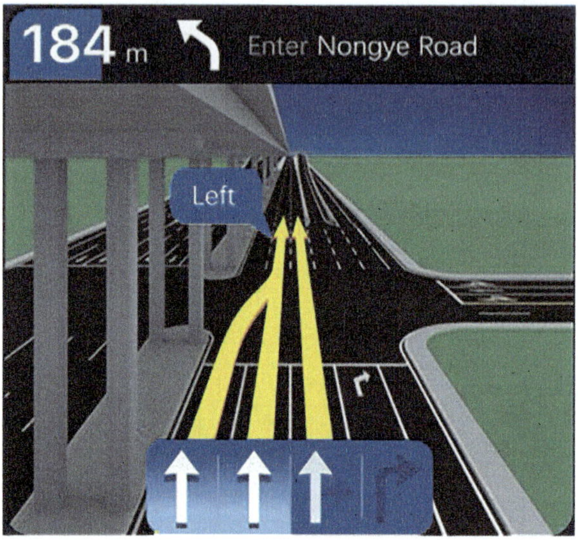

Fig. 8.9 Countdown progress bar in the Amap navigation system (2021), translated from Chinese language

does not provide more information but rather makes the existing information easier to understand. This is a notable difference from the availability index under utility.

Cross-reality operation enables users to directly operate the functions of a vehicle on a realistic graphical interface rather than selecting options from information lists or matrices. Cross-reality operations not only make interactions more intuitive and easily understandable but also enable the screen interface to facilitate the synchronized mapping of the actual vehicle in the physical world to its digital counterpart, thereby enhancing the sense of technology. For example, as shown in Fig. 8.11, the trunk lid of this virtual vehicle can be directly dragged by the user to synchronously control the opening angle of the actual trunk lid.

Interactability Recognition

In automotive HMI systems, screens often display several elements simultaneously, and users need to know which elements are tappable and which are not. For lists or matrix-style menus, the interactable elements are usually neatly arranged and easily identifiable. However, for cross-reality operation interfaces, the layout of interactable elements is irregular, which implies that these elements need to be represented using special graphics, borders, colors, or animations to make them easily identifiable to

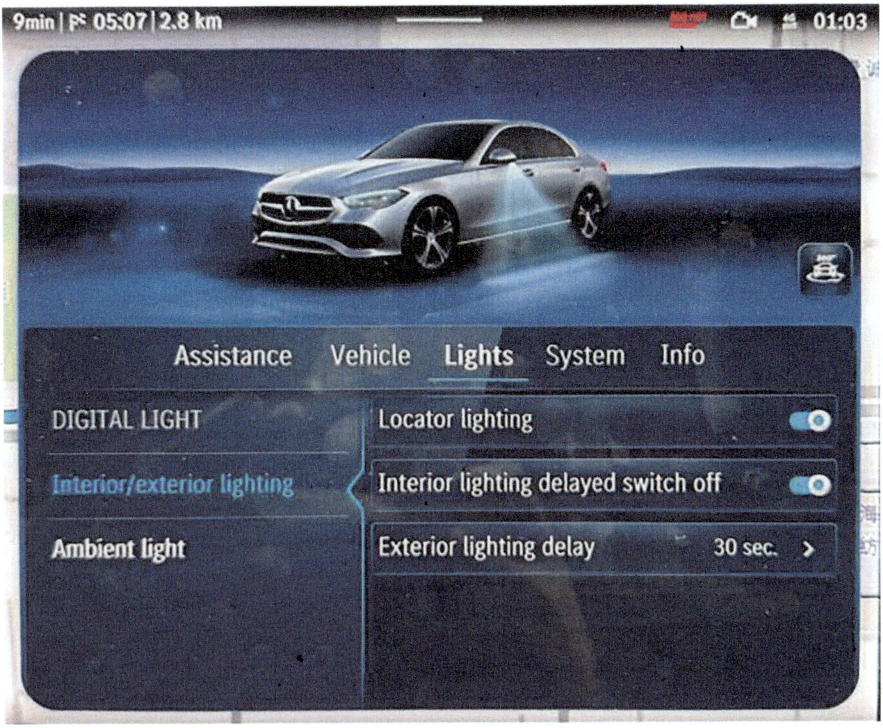

Fig. 8.10 Visual interface for light settings in Mercedes-Benz C-Class (2023)

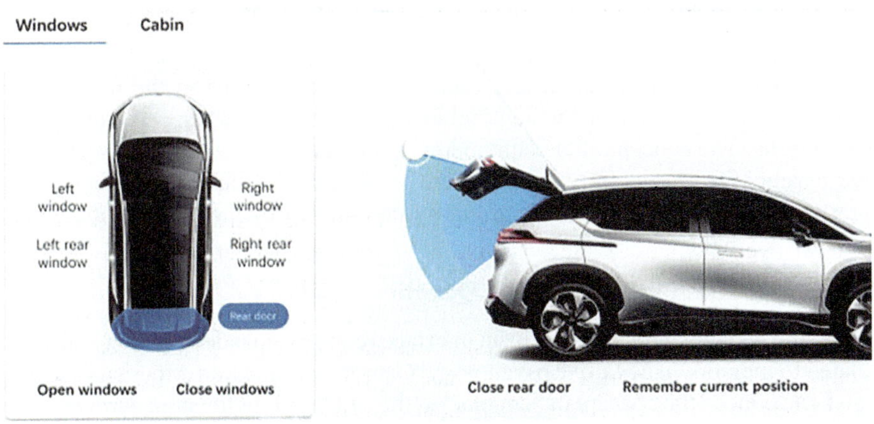

Fig. 8.11 Direct operation of the opening angle of the trunk lid in GAC Avion V Plus (2021), translated from Chinese language

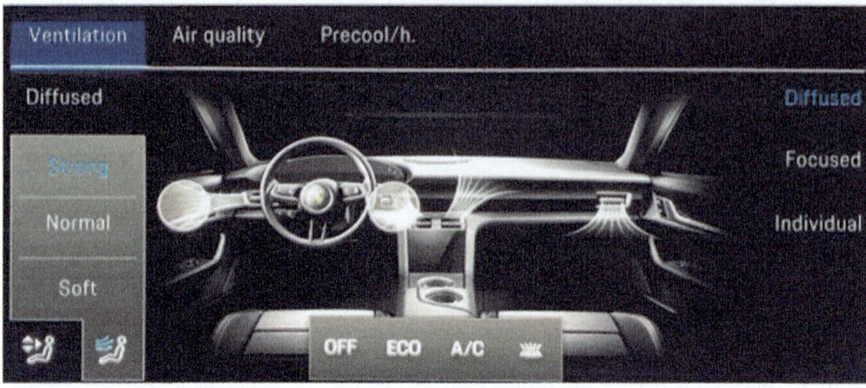

Fig. 8.12 Climate control interface of Porsche Taycan (2020), wherein users can adjust the airflow direction by dragging the circles on the two left-side air vents with fingers, translated from Chinese language

users. For example, in the cross-reality operation interface of the climate control system shown in Fig. 8.12, the two left-side air vents have semi-transparent white-bordered circles in the front of their wind direction, indicating that they can be dragged, whereas no such graphics are present for the right-side air vents, indicating that they cannot be dragged.

8.2.4 Element Memorability

Element memorability refers to the extent to which elements in the interface are arranged in a simple and easily memorable manner for users when performing interaction tasks. The concept of element memorability is closely associated with cognitive psychology, particularly memory and thinking. Whether the elements in the interface are easy to remember is primarily determined by the number of elements and their layout.

Number of Icons/Phrases and Text Quantity

Assessing element memorability in an interface requires an understanding of theories related to memory in cognitive psychology. George A. Miller, an American psychologist, introduced the concept of "chunks" as the smallest unit for short-term memory, which is a unit that people are familiar with. Graphical chunks typically consist of a single graphic or a group of graphics, whereas textual chunks can encompass a character, letter, word group, or phrase. For instance, the phrase "Technical University of Munich" is considered one chunk because users tend to perceive it as an indivisible whole during the memorization process. Conversely, the phrase "Technical Munich" is recognized as two chunks despite having only two words owing to the lack of a strong association between these two words in most contexts, thus requiring users to memorize them separately.

In a specific page on the screen, all elements can be classified into several primary chunks with respect to graphic borders as well as icon layout and clusters. For example, in Fig. 8.13, chunks 2, 3, 4, and 5 are clearly delineated by graphic borders that outline their boundaries. However, although chunks 1 and 6 do not have explicit borders, and the icons within them are neatly arranged and regularly distributed, which shows a prominent unity. Each primary chunk can be further divided into several secondary chunks. For instance, in the primary media chunk (chunk 4), some secondary chunks are present, including the play icon, next track icon, favorite icon, digital radio icon, and channel title. When users memorize the position of the "favorite" icon, they do not directly remember its location on the entire screen. Instead, they first remember that it is in the media module in the lower-left corner of the screen and then memorize its position in the lower-left within the media module.

To facilitate users' memorization, the number of both primary and secondary chunks should be limited. The appropriate number of chunks at each level can be guided by research in cognitive psychology. For instance, George A. Miller found that the short-term memory capacity of individuals is approximately 7 ± 2 chunks [1].

Secondary chunks often include continuous text, which can be a single word or a phrase. These continuous texts should not be too long or they may hinder the users' memorization and comprehension. However, there are two types of texts whose length can be disregarded. The first is open-ended content entered by the user, such as song titles or destination addresses, as their length is not under the control of the HMI system designer. The second is the descriptive or explanatory text for specific functions; once users understand these functions, they no longer need to carefully read the associated explanatory text in subsequent use.

Layout Method

During the process of memorizing and searching for chunks, people typically follow certain patterns or sequences, such as scanning row by row, column by column, or in a clockwise manner. Therefore, the core chunks on the screen should also have a clear and coherent layout. For instance, icons on the main menu can be arranged in a row, a column, a two-dimensional matrix, or even in an arc. Conversely, irregular layouts, such as diamond-shaped, honeycomb, random bubble, or word cloud layouts, are not conducive to users' memorization and searching.

8.2.5 System Feedback

When performing interaction tasks, the system should be able to provide users with prominent, immediate, and smooth feedback after receiving their input information. System feedback is closely associated with perception and cognition in cognitive

Fig. 8.13 Division of chunks on the homepage of the central information display of the Nio EL7 (2023)

psychology. Effective system feedback not only assists users in completing interaction tasks more easily, naturally, and efficiently but also may shape their overall subjective impression of an HMI system.

Feedback Richness

System feedback in HMI should occur at two stages: input success and execution success. Input success feedback should occur when the input is completed, informing the user that their input was successful. Execution success feedback is intended to show the user the execution result and prompt them to perform the next step. When the system is running smoothly, input and execution success may occasionally occur almost simultaneously. However, we should neither confound the two nor arrange all feedback to occur only at the moment of execution success. When the system experiences delays, there will be a time gap between the input and execution success. If the system fails to provide timely feedback after input success, users may repeat their operations, thus resulting in a longer duration of distraction and even interaction path errors that require backtracking and reinitiating the task. Each interaction modality has its corresponding method of system feedback, as presented in Table 8.3.

As a traditional interaction modality, physical buttons can transmit the input success feedback to the finger through haptic sensations, such as the stroke of pressing the button or the stroke of rotating the knob. This is a notable advantage of physical buttons. By contrast, touch-sensitive buttons do not have a stroke; hence, they require the vibration of touch panels to deliver feedback to the user's fingers. Additionally, after pressing a button, the system can also emit a brief auditory prompt. For buttons that are not linked to the in-screen information, execution success is typically indicated through changes in the button's appearance or backlight. In the case of buttons linked to the in-screen information, execution success is often indicated by the appearance of pop-up windows on the screen or changes in the on-page information.

Touchscreens lack haptic feedback owing to the absence of a pressure stroke, which is a disadvantage compared with physical buttons. To compensate for this, some in-vehicle touchscreens are equipped with haptic feedback that resemble the vibration sensation of touch buttons. However, owing to high manufacturing costs, vibration feedback on touchscreens has been slow in gaining popularity. The icons or text on the touchscreen change their color and shape when they are successfully tapped. For example, in the Apple iOS operating system, when a user taps on an icon on home page, the icon first turns gray and then enters the corresponding application, as shown in Fig. 8.14. These changes may be subtle and not explicitly noticeable to

Table 8.3 Suggested system feedback for each type of interaction modality

Interaction modality	Input success feedback	Execution success feedback
Button	• Haptic sensation (e.g., vibration and button stroke) • Auditory prompts	• Changes in the button appearance • Changes in the backlight • Pop-up window/changes in the on-page information
Touchscreen	• Auditory prompts • Changes in the icon/text color or shape • Screen vibration upon touch	• Pop-up window/changes in the on-page information
Voice	• Voice confirmation (including repeating instructions) • Displaying command text on the screen	• Pop-up window/changes in the on-page information • Voice notification of the execution result

Fig. 8.14 Icons in the Apple iOS 15 smartphone system (left) turn gray after being tapped (right)

users, but they contribute to a sense of smooth, rhythmic, and responsive interaction, enhancing the overall user experience.

The interaction modality of voice control differs substantially from that of buttons and touchscreens, as its input success feedback typically relies on the response from the voice system. For example, the system may confirm the user's command with a response such as "okay" or repeat the user's command by saying "are you trying to navigate to xxx?" In addition, it is beneficial if the voice control system displays the user's command on the screen in real time as well. This not only demonstrates that the voice control system is listening and comprehending but also allows the user to verify whether their speech has been correctly recognized. If a recognition error exists, the user can promptly pause and repeat the command as early as possible. For the output success feedback, voice control should not only display pop-up windows or page changes on the screen but also inform the user through system responses, such as "the temperature has been set to 26°." However, excessively lengthy system responses can potentially increase the task duration and cognitive distraction.

Perceptual Smoothness

The smoothness of touchscreen responsiveness often forms a user's initial impression of an automotive HMI system. Although a brief momentary lag may not necessarily impact efficiency and safety, it can adversely affect the user's overall subjective impression of the system. Achieving excellent perceptual smoothness requires a minimal time interval between the end of user operation and perceived system feedback as well as natural and fluid feedback animations. Tasks that significantly affect perceptual smoothness include swiping between pages on the screen quickly and freely or tapping icons in the main menu to access specific applications.

Eliminating Waiting Anxiety

Occasionally, an automotive HMI system may take a few seconds to process information when performing certain tasks such as recognizing the user's voice input or generating navigation destinations based on keywords entered by the user. These information processing tasks involve substantial computational load and often rely on support from cloud-based resources, making it challenging to reduce the processing time. During the waiting period, users may perceive a disruption in the interaction process and even suspect a system malfunction, leading to anxiety. In such cases, the system should use on-screen text, animations, or voice prompts to inform users that the system is currently processing the information to alleviate their anxiety during the waiting time. If these animations and voice prompts are made lively and interesting, they may further narrow the emotional distance between people and vehicles.

8.3 Summary of the Evaluation Results and Design Suggestions

8.3.1 Trends in the Interaction Hierarchical Structure

As a reference for automotive HMI systems, the interaction hierarchical structure of cell phones has evolved constantly over the past two decades. In functional phones, the home screen typically consists of a wallpaper and two shortcut icons, requiring one tap to access the main menu. In comparison, the root directory layer has been transformed into the home screen in smartphones, which includes not only the icons of commonly used applications but also frequently accessed toggles of settings and some application widgets. The hierarchical structure of automotive HMI systems is more diverse and less uniform than that of smartphones. For example, some car models have a homepage that is the root directory, whereas some models feature a homepage that is a shortcut layer.

The HMI system of several traditional automotive brands uses the root directory as the homepage. Figure 8.15 shows the evolution of the iDrive system in the BMW 5-Series over the past 20 years. From 2003 to 2015, the homepage of the BMW 5-Series was a very typical root directory, allowing access to all functions from this page, except for the fact that the root directory was distributed in four directions in the E60 model and was a vertical list in the F10 model. In 2016, functional card-style widgets were introduced in the G30 model, with six cards spread across two pages. These six cards continued to serve as the root directory, as they provided access to all functions. In 2020, the G30 facelift model underwent significant changes, with most of the home screen occupied by customizable functional cards, while the root directory, which provided access to all functions, became five icons arranged vertically on the left edge. BMW's persistence in using a root directory–style homepage and a logic tree–structured menu is driven by both historical continuity and considerations for the usage characteristics of the iDrive rotary knob. The rotary knob allows for natural horizontal or vertical movement of the cursor but, unlike touch input, lacks the freedom of selecting arbitrary coordinates on the screen.

Fig. 8.15 Homepage layout of successive generations of the BMW 5-Series iDrive system (*Source* BMW Group)

In recent years, there has been a notable trend in the hierarchical design of automotive HMI, wherein the shortcut layer is used as the home screen instead of the root directory layer. The shortcut layer can be used as the home screen via two common approaches: the first approach is to arrange several widgets on the home screen, as shown in Fig. 8.16. These widgets primarily include map widget that display routes and guidance information as well as music widget that show the current song track and allow switching to the next track. The number and placement of these widgets are typically customizable. Tapping on a widget can also take the user to the full-functionality application page, that is, the widget is an entry point. However, most car models do not allow access to full-functionality applications through widgets; therefore, widgets are not a substitute for the root directory. The second approach is to use a complete map as the home screen, as shown in Fig. 8.17. In this case, the map serves not only as an application but also as the shortcut layer. This design integrates navigation, points of interest search, and other ecological functions into a single map, which emphasizes product positioning for urban exploration. Designs that use a map as the home screen usually require an additional one or two parallel home screens or two to four widgets on the home screen, because not all functions can be accessed from the map; for example, music player is unrelated to the geographic information displayed on the map. Many car models that use widgets or maps as the home screen also provide an additional application list page, which serves as the root directory but is not frequently used.

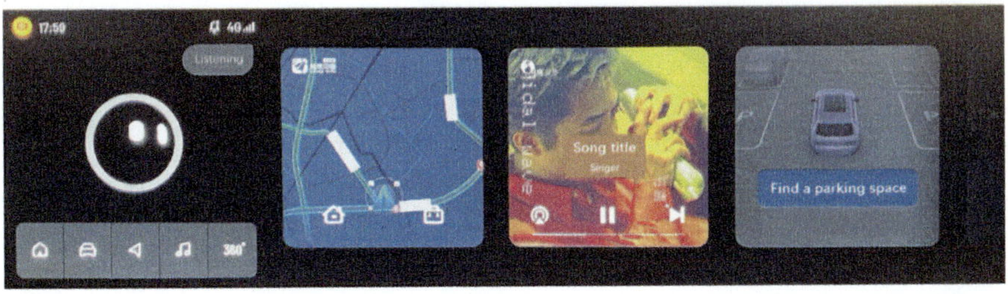

Fig. 8.16 Widgets on the central information display of Li Auto One (2021), translated from Chinese language

Fig. 8.17 Central information display using a map as the home screen with several widgets in GAC Aion V Plus (2021), translated from Chinese language

Theoretically speaking, there is no distinction in the superiority of different homepage types or hierarchical structures as long as the design can minimize the operation steps, provide intuitive information, and maintain a clear logic and is easy to learn and use. When evaluating actual car models, products with excellent designs for an interaction hierarchical structure mostly use widgets or maps as the home screen. Although traditional designs that use the root directory as the home screen are easy to learn, they often lack potential in terms of intuitive information display and operation step reduction.

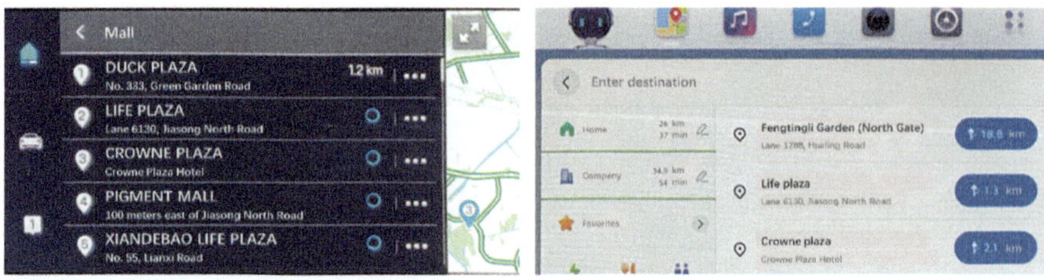

Fig. 8.18 Font size comparison between Porsche Taycan (2021, left) and Xpeng P5 (2021, right), translated from Chinese language

8.3.2 Element Size and Esthetics

In practice, larger screens do not necessarily entail larger icons and text and hence better visibility. A comparison between the 2021 Porsche Taycan and Xpeng P5 revealed that the former has a smaller screen size of 10.9 inches with an average font height of 5.0 mm for its main functions, whereas the latter has a larger screen size of 15.6 inches but an average font height of only 3.4 mm, as shown in Fig. 8.18.

The number of car models incorporating small-sized elements into large-sized screens is increasing. This is partly due to increased screen resolution and definition, which allows smaller elements to be displayed more clearly and sharply. Additionally, it enhances the esthetics of the page layout. As the element size is reduced and white space is increased, the interface appears cleaner and simpler while also exuding a sense of technology. This design approach is not only applicable to vehicle interfaces but is also widely used in the design of smartphones, tablets, and other consumer products. For most page layout styles, there is a trade-off between element visibility and esthetics. Therefore, balancing and making compromises in this regard require designers to consider brand positioning and the target group.

8.3.3 Does Learnability Matter?

Learnability is an important component of the usability of interaction systems. Items 3 and 4 in the standard SUS questionnaire specifically address learnability, as presented in Table 8.2. The factors that influence learnability are relatively complex and comprehensive, involving indexes such as logical structure, understandability, and memorability. However, learnability also extends beyond the scope of these indexes and is difficult to quantify; therefore, it is not included as an evaluation index in this chapter.

Is learnability important for automotive HMI systems? Some argue that it is because a complex interaction system lowers the learning threshold for users. Others argue that it is not as important because users typically drive the same vehicle for several years. Thus, even if the initial learning curve is steep, it does not affect the long-term user experience. There is no definite answer available in this regard; nevertheless, the following points should be considered: first, the baseline for learnability is the existing usage habits of the target users, which can vary among target groups. Products targeting younger users can introduce innovative designs for them to explore and learn as long as the designs are logically reasonable. Inspiration can also be drawn from popular smartphone applications. Products targeting older users should maintain a high level of consistency with previous generations, minimizing significant changes in interaction methods, even if the new interactions may be more reasonable. Second, unique new functions should be easily discoverable by users; otherwise, some users may remain unaware of these functions for a long time. For example, since 2015, some German brand models have integrated applications popular in China (e.g., Weibo and Dianping) in vehicles; however, they are often hidden deep within the menu. Consequently, finding them is often difficult unless the users actively explore all functions. Third, the level of learnability that users are willing to accept is inversely proportional to the dominance of the brand. Users' attitudes towards the interaction experience are not purely rational but instead can be influenced by emotion. When facing a car model from a dominant brand, many users believe that they should invest time in learning its interaction methods. However, users are more likely to lose patience and abandon the learning process in car models from less-dominant brands. For example, Tesla Model 3, launched in 2017, completely consolidated instrument cluster functions into the central information display, which disrupted users' previous usage habits; yet, it was widely accepted. In the next five years, few brands had dared to try a similar integration.

References

1. Herbert A·Simon. Cognition: The thinking and intelligence behind human behavior [M]. Beijing: Renmin University of China Press, 2020.
2. Margaret W.Matlin. Cognitive Psychology, 8th Edition [M]. Translated by Li Yongna. Beijing: China Machine Press, 2016.
3. Goldstein, E. B. Cognitive Psychology, 3rd Edition [M]. Translated by Zhang Ming. Beijing: China Light Industry Press, 2015.
4. U.S. Department of Transportation. In-Vehicle Display Icons and Other Information Elements Volume I: Guidelines [R]. 2004.
5. NHTSA. Human Factors Design Guidance For Driver-Vehicle Interfaces [R]. 2016.
6. Road vehicles — Symbols for controls, indicators and tell-tales: ISO 2575: 2021 [S].

Chapter 9
Intelligence

9.1 Development

9.1.1 Definition of Intelligence

Intelligence refers to the aggregate or global capacity of an individual to act purposefully, think rationally, and manage effectively with their environment. This definition was proposed by the renowned psychologist David Wechsler in 1944 in his book *The Measurement of Adult Intelligence* [1]. Intelligence can be classified into human intelligence, nonhuman animal intelligence, and artificial intelligence (AI).

AI refers to the ability of a system to perceive the environment and perform actions to achieve specified objectives to the greatest extent possible. The evaluation criteria for AI or machine intelligence are similar to those for human intelligence, which involve taking dynamic environmental changes as the system input, implementing calculations and processing, and generating more efficient outputs to achieve specified objectives. In this definition, manually operated windshield wipers are not considered an intelligent system even if the switch is on the central information display. By contrast, windshield wipers that can automatically sense rainfall and activate themselves are considered an intelligent system even if they also have a traditional physical lever.

It should be noted that the term AI used here is broader than the narrow sense of AI. AI is a specific research field that has emerged in recent years and generally involves the use of machine learning algorithms for tasks that cannot be handled by simple causal rules. However, for the AI definition provided in this chapter, it is not necessary for the system to use complex algorithms.

In addition to intelligence, "smart" has similar meaning sometimes. Smartphones and smartwatches are commonly used instead of intelligent phones or intelligent watches. "Smart" does not constitute the scope in academic research but rather is a characteristic of consumer products. Although a smart device has no strict definition,

J. Ma and Z. Gong, *Automotive Human-Machine Interaction (HMI) Evaluation Method*, Research on Automotive Intelligent Cockpit,
https://doi.org/10.1007/978-981-99-9951-4_9

it typically encompasses three features: powerful computing capabilities, real-time access to the Internet, and an open operating system.

For machines, intelligence and smart have similar meanings which often overlap but are not equivalent. Intelligence places more emphasis on the ability to handle each specific task. For example, brands such as Mercedes-Benz and Nissan have "intelligent headlights" to emphasize their ability to provide different lighting solutions under various environmental conditions. However, headlights rarely have new functions other than illumination. Conversely, smart emphasizes the diversity of tasks that can be handled, like "Smart TV", and the most significant difference between smart and traditional televisions is the availability of richer content resources.

"Intelligent vehicles" is more popular than "smart vehicles" or "smart cars." The reason may be that, in addition to the intelligent cockpit HMI system, intelligent vehicles also encompass autonomous driving, which is clearly beyond the definition of a smart device. In terms of the definition scope, if we only discuss intelligent cockpit HMI systems without involving autonomous driving, the use of the term "smart" would not be inappropriate, however, most people are still accustomed to using "intelligent."

Automotive HMI combines both meanings of intelligence and smart. On one hand, the automotive cockpit is becoming increasingly similar to a smart device. Features such as videos, games, and lifestyle services have gradually been incorporated into the vehicle. In fact, many automotive HMI systems even utilize the Android operating system, which is similar to that widely used in smartphones and televisions. On the other hand, an automotive HMI system is not simply a collection of functions. It needs to collect real-time positioning, driver status, and other traffic information as well as several other environmental factors to provide more proactive and efficient services to users. We refer to these two directions of automotive HMI development as functional and contextual intelligence, respectively.

9.1.2 Automotive Cockpit as the Best Carrier of Intelligence

Automotive HMI is expected to become the most intelligent device available to consumers in the future. The automotive cockpit will also become the most intelligent space that consumers can access. These statements may sound radical, especially given that the current level of intelligence in automotive HMI systems is not yet on par with that in smartphones. However, when we strip down both automotive HMI systems and smartphones to machines for computation and analyze their architectures, we will discover the advantages of automotive HMI.

All computational machines require input and output devices. In 1949, John von Neumann, "the Father of the Modern Computer," proposed a computer architecture that consists of five components: input devices, memory unit, arithmetic/logic unit, control unit, and output devices, as shown in Fig. 9.1. Input and output devices are not unique to electronic computers. As early as 1833, the British inventor Charles

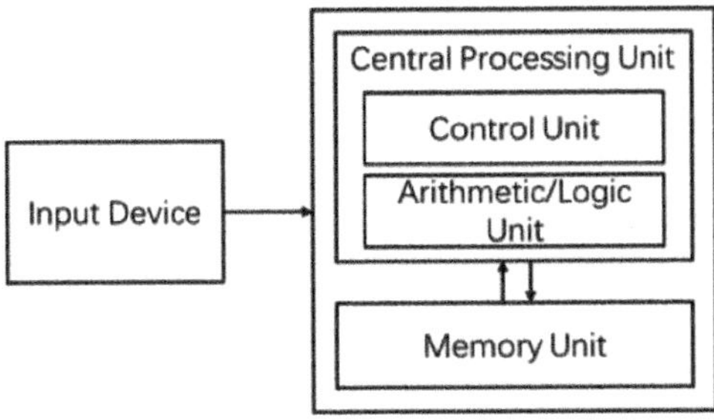

Fig. 9.1 Computer architecture proposed by John von Neumann

Babbage incorporated punched cards as input devices in his Analytical Engine (a purely mechanical calculator) and used a printer, plotter, and bell as output devices.

The intelligence level of a machine relies heavily on its input and output devices. According to David Wechsler's definition of intelligence, the machine should be able to adapt to specified environments, that is, it should receive sufficiently rich environmental information as an input. Simultaneously, the machine should be capable of efficient response and execution, which requires it to provide sufficiently rich forms of output. These inputs and outputs should be highly automated, minimizing the need for human intervention. Therefore, sensors play a crucial role as input devices, whereas actuators are essential output devices.

Although smartphones have powerful computational capabilities, they have a limited number of built-in sensors, including cameras, microphones, and inertial measurement units (IMUs). Moreover, some sensors may not work effectively when the smartphone is not in use. For example, when a smartphone is placed face up on a flat surface, the rear camera is unable to capture anything and the front camera can only capture the ceiling, thus providing no valid information. By contrast, the cockpit of an intelligent vehicle can be equipped with a wider variety of sensors that can work effectively as long as the vehicle is in use. For instance, in the cabin, seat cushion sensors can detect the presence of passengers, in-vehicle cameras can monitor user actions and expressions, steering angle sensors can determine driver fatigue, and microphone arrays can not only capture user voices but also locate which specific user is speaking. On the vehicle body and chassis, wheel speed sensors combined with IMUs can precisely determine the vehicle's state during motion, and the power system can monitor real-time energy consumption. For the external environment, cameras and radar can monitor surrounding vehicles, pedestrians, and obstacles, while rain and light sensors can assess current weather conditions.

The output devices of smartphones are even more limited, typically consisting of only the screen and speaker. Additionally, the screen and speaker can neither directly impact the motion of any physical device nor do they fall under the narrow definition of actuators. By contrast, vehicles have a wide range of actuators in addition to screens and speakers. Each electric machine in the vehicle can be considered an actuator, allowing adjustments in the seat position, mirror angle, trunk lid opening/closing, climate control airflow, and windshield wiper speed. Moreover, the vehicle itself is a large-scale actuator as it can maintain safe driving by controlling the outputs of the powertrain and chassis systems, avoiding loss of control and collisions with other vehicles, pedestrians, and obstacles. With advanced autonomous driving, the entire vehicle can also serve as a mode of transportation, acting as an actuator that transports users to their destinations.

The abundant sensors and actuators in vehicles are designed, calibrated, and managed in a unified manner, enabling a high level of synergy among them. Although smartphones can connect to some smart home devices to access a wide variety of sensors and actuators, such user-configured systems cannot achieve the same level of coordination as that observed in vehicles.

The rich array of sensors and actuators in vehicles can create vast space for potential intelligent scenarios. Once the user enters the vehicle, the vehicle can automatically adjust the climate temperature and ambient lighting color according to the driver's physical and mental state. The vehicle can also automatically adjust the seat position and rearview mirror angle based on the current sitting posture. After the navigation destination is set, the vehicle can automatically detect gas stations or restaurants along the route and recommend 2–3 meal options according to the user's preferences for quick selection. During the driving process, the vehicle can automatically adjust the brightness, range, and beam shape of the headlights based on environmental factors such as lighting intensity, position of nearby vehicles, and weather conditions. Upon arrival at the destination, the vehicle can drive autonomously from the building entrance to the designated parking space without the need for driver control. To realize these intelligent scenarios, the automotive HMI system must be made the core of computation and decision-making, while all input and output devices in the vehicle should be fully integrated. However, if the relatively isolated smartphone is considered as the core instead, it will be challenging to integrate these vast amounts of data.

Vehicles have tremendous potential for intelligent scenarios, but realizing this vision is not an easy task. First, the traditional electrical/electronic architecture of vehicles isolates the various systems from each other. For instance, although vehicles can collect real-time wheel speed data, these data may not be fed to the navigation system to correct real-time positioning. Connecting these data requires not only scenario definitions based on user experience but also an upgrade of the electrical/electronic architecture of the entire vehicle. Second, software and algorithms are not the strengths of traditional automotive companies. Once all the data are interconnected, powerful software and algorithms are needed to analyze them. The automotive industry generally lags behind the Internet industry in this regard, which has limited the output of intelligent experiences. For example, the concept of recommending 2–3 meal options based on user preferences may sound simple; however, in reality, there is currently no smartphone software that can perform this task satisfactorily, and most users still spend a considerable amount of time browsing through lengthy menus. Additionally, data security and privacy protection are crucial issues. Even if various data can be fully integrated at the engineering level, complying with the laws and regulations related to data security is still essential. Owing to the rapid development of the intelligent automotive industry, there is often a lag in the formulation of relevant laws and regulations.

9.2 Evaluation Indexes

The evaluation of automotive HMI intelligence can be divided into second-level evaluation indexes, which include comprehension, functional intelligence, and contextual intelligence.

9.2.1 Comprehension

Comprehension refers to the system's ability to understand the users' natural commands and engage in effective interactions. For automotive HMI systems, comprehension primarily focuses on the voice control modality. The voice commands from users are often not pre-defined words but rather colloquial sentences with contextual logic. Therefore, in addition to recognizing each word, the voice interaction system needs to analyze and understand these words to fully grasp the user's true intention. By contrast, interaction tasks using modalities such as touchscreens or buttons have clear operational purposes and a limited range of choices, eliminating the need for the system to comprehend the user's input. For example, in a list of navigation destinations on the central information display, each page may present six options, and the user can select and tap on one of these options. Subsequently, the system can accurately determine which option the user has selected based on the coordinate value of the touch point on the screen. In the future, the proliferation of more natural HMI modalities, such as gesture interaction, facial expression interaction, and brain-machine interfaces, may further expand the application scope of comprehension. Intelligent functions that do not require active user input do not fall under the scope of comprehension and will be discussed under contextual intelligence. Although excellent comprehension capabilities can make interactions more natural and convenient, they do not increase the number of functions or directly improve task success rates. Therefore, comprehension should not be confused with indexes under utility.

The in-vehicle voice interaction system should communicate freely with the users within a specified scope of objectives, providing them with a sense of efficiency, convenience, authenticity, reliability, and respectfulness. As voice control offers great flexibility in both input and output, designers can utilize distinctive responses to imbue vehicles with specific personalities, thereby enhancing the human–vehicle relationship, and enabling the vehicle to become more than just a tool.

The foundation of voice control comprehension lies in achieving natural language conversations [2]. Early voice control systems could only recognize specified mechanical commands such as "raise the temperature". These commands could not be modified by users as any changes may impede system recognition. Natural language conversations significantly relax the constraints on the range of commands, allowing users to express themselves similarly to that in normal interpersonal communication. For example, phrases such as "It is too cold" or "I feel a bit chilly" can be understood as a request to raise the temperature of climate control.

In addition to natural language conversations, voice control should also possess comprehension capabilities such as contextual understanding, interruptibility, error correction, and sound-source localization. Contextual understanding refers to the system's ability to infer the topic of discussion based on the context when users engage in continuous conversation for more than one round with the HMI system. For example, if a user asks about the weather in Munich and then follows up with "what about Stuttgart?", this indicates their interest in knowing the weather condition in Stuttgart rather than seeking other information about the city. Interruptibility means that users can interrupt the system's voice announcement and directly state the next command if they have already understood the system's intention before it completes the full announcement. This significantly improves the efficiency of the voice interaction process. Error correction allows users to rectify partial information when they make a mistake in their expression, without the need to restart the conversation. For instance, if a user incorrectly states the 10th digit of an 11-digit phone number, they can simply repeat the last four digits, avoiding the need to repeat the entire number. Sound-source localization refers to the ability of the voice interaction system to recognize which occupant in the vehicle is speaking via directional microphones. For example, upon identifying that a rear passenger has said "raise the temperature," the system can adjust the temperature specifically for the rear seats.

9.2.2 Functional Intelligence

Functional intelligence in automotive HMI systems refers to the quantity and richness of open-ended applications that are not directly related to driving and vehicle control. These applications typically require Internet connectivity and may include entertainment applications, such as music, videos, and games, as well as service applications, such as dining, car wash, and parking.

The openness of these applications is reflected in two aspects: first, users can access newer and more applications through online downloads and upgrades. Second, the content of these applications is updated in real-time from online servers rather than being fixed within the local system. These applications can assume various forms within the automotive HMI system. They can exist as standalone software applications with their icons serving as gateways, similar to applications on smartphones. Alternatively, they can be integrated into existing modules. For example, users can choose to access an online music library from the music playback interface or they can click on the restaurants that appear on the map to make reservations.

Function Richness

Function richness refers to the number of functions covered by open-ended applications in automotive HMI systems. In the actual evaluation process, a function library can be established, and the proportion of functions provided by a specific HMI system can be examined.

The purpose of automotive functional intelligence evaluation is to integrate more valuable applications into the automotive HMI system and provide a better user experience. Therefore, in the evaluation process, we can consider excluding implementation methods for applications that do not fully fall under the capabilities of the automotive HMI system. Three types of methods can be excluded: first, methods using remote human customer service to implement specific types of functions or services such as General Motors' OnStar. Although human services can provide many functions, the automotive HMI system primarily serves as a communication device throughout this service process, with virtually no involvement of its own intelligence. Second, methods using third-party non-native in-vehicle devices to implement specific types of functions or services such as the Apple CarPlay projection. CarPlay is only supported by Apple devices but not Android devices and hence does not have user universality. Additionally, the automotive HMI system only serves as an output device and does not independently provide any intelligent services. Third, methods using non-automotive scenario applications to implement specific types of functions or services such as logging into the web version of WeChat through a browser on the central information display. Browser-based services are not optimized for driving scenarios and usually provide a poor user experience. Moreover, if an automotive HMI system can achieve the majority of functions and services simply by having a browser, further evaluation of such systems would be of limited significance. Common open-ended functions and services in automotive HMI systems are presented in Table 9.1.

Occasionally, an automotive HMI system may provide multiple applications with similar functions. For example, in the Chinese market, some vehicles offer various online music applications, including QQ Music, Kugou Music, and Tingban, or various map navigation applications, including Amap, Baidu Map, and Tencent Map. From the perspective of function richness alone, having more applications of the same type is considered better as it provides users with more choices. However, if

Table 9.1 Common open-ended functions and services in automotive HMI systems

Lifestyle services	Restaurant reservations, smart home devices, photo taking, photo album, vehicle maintenance, gas/charging payment, parking payment, and shopping
Entertainment	Online music, online audio, videos, games, and karaoke
Navigation ecology	Account sharing with mobile maps, location pushing/sharing with other applications, proactive destination recommendations, point-of-interest analysis, records of frequently used routes, and road trips
Social ecology	Instant messaging, group travel, and stranger socialization
News information	News and weather

we consider the overall user experience, having too many applications of the same type can overwhelm users and make it difficult for them to choose. A better approach is for the automotive HMI system to integrate the resources of all similar applications into a unified platform. For instance, when a user searches for a specific song on a single online music platform, the system can automatically search for the highest quality version from multiple online music applications, eliminating the need for users to make inefficient choices, judgments, and corrections.

The applications provided by some car models are not directly sourced from Internet companies but instead involve the participation of automotive manufacturers. These applications can generally be categorized into three types. The first type is a platform-type application that integrates content resources from multiple applications. The second type uses the brand's DNA to filter content. For example, a luxury automotive brand can package recommendations of high-end restaurants that align with its brand tonality in a restaurant recommendation application. The third type is the brand own created content. For instance, Nio Radio, an online radio station by Nio Inc., is the world's first user-created audio community for Internet-connected vehicles.

Whether having more open-ended functions in automotive HMI systems is better remains a topic with no absolute consensus in the industry. Why would users operate the vehicle's central information display to book a train ticket instead of booking it directly on their smartphones? Similar questions arise because the convenience of using these applications on the central information display does not necessarily surpass that of using a smartphone directly. However, if in addition to purchasing a train ticket, parking applications can automatically reserve a parking space at the train station for the user or if the ticketing application can provide rescheduling recommendations when encountering severe traffic congestion on the way to the train station, the advantages of the automotive HMI system over a smartphone become more apparent. Thus, several seemingly redundant in-vehicle functions are not actually useless; it means that the current design has not yet fully optimized the user experience flow.

In some standardized evaluation processes, where it is not feasible to fully and quantitatively assess the functions provided by each application, we can make the following general assumptions: it is better to have more types of functions and services; it is better to have a greater number of applications within each type; and brand-customized applications are superior to generic applications.

Content Resource Richness

Content resource richness refers to whether the online resources provided by open-ended applications in automotive HMI systems can meet the users' common needs. These online resources include purely digital content resources (e.g., music and movies) and points of interest with real locations (e.g., restaurants and gas stations).

When evaluating content resource richness, the content contained in third-party non-native in-vehicle devices, such as the Apple CarPlay projection, is typically not included. Under current technological conditions, smartphones serve as the gateway to almost all online content resources. For example, nearly any song can be found

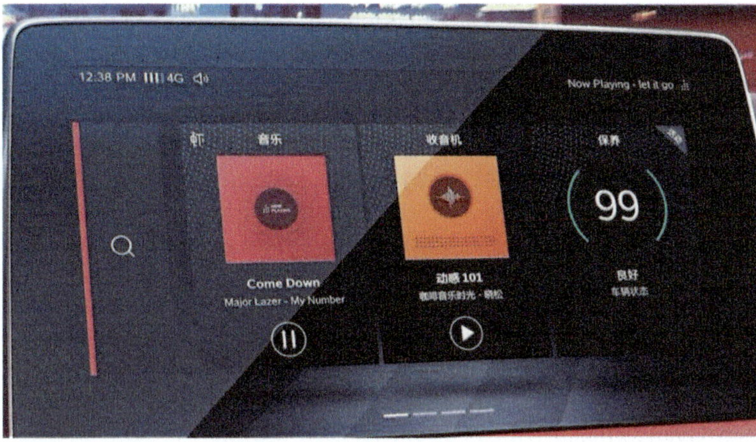

Fig. 9.2 Banma Intelligent HMI System offering Xiami Music in MG HS (2018) (*Source* SAIC MG)

in mainstream online music applications, and almost all points of interest, including restaurants and gas/charging stations, can be found in mainstream map or service applications. If all these content resources from smartphones are considered, all vehicles would have access to highly comprehensive content resources, rendering the evaluation meaningless.

Content resource richness is subject to dynamic changes. For example, in the Chinese market in 2017, SAIC Roewe RX5 was launched, which was equipped with the Banma intelligent HMI system jointly developed by the SAIC Group and Alibaba, and integrated with the then-rich resource platform Xiami Music. This system was later adopted in other car models under the SAIC Group, including MG HS, as shown in Fig. 9.2. However, owing to Alibaba's gradual defeat in the competition for music copyrights against Tencent and NetEase, music resources in the Banma system decreased significantly. Subsequently, the Banma system introduced music resources from the Tencent-owned platform to expand its online music resources.

9.2.3 Contextual Intelligence

Even if an automotive HMI system has a wide range of functions and abundant content resources, this does not necessarily ensure good usability. The functions and content in HMI systems should also be matched and optimized for in-vehicle scenarios to achieve better contextual intelligence.

Contextual intelligence is more important for vehicles than for smartphones. This is because drivers often need to operate the automotive HMI system while driving, and significant driver distraction can cause potential hazards in driving safety. Therefore, intelligent functions should neither consume excessive time nor require substantial effort. For example, on a smartphone, users often spend several minutes browsing a restaurant menu, making selections, and placing an order; however, in a driving scenario, even a few seconds spent browsing the menu may pose serious safety hazards. Therefore, food service applications need to recommend a very limited number of choices to the users to minimize driver distraction. Such accurate dish recommendations rely heavily on rich data collection and powerful intelligent algorithms.

Furthermore, in non-driving scenarios, smartphone users have moderately low cross-application demands, whereas such demands occur very frequently during in-vehicle driving scenarios. For example, when using food service applications on a smartphone at home or in a shopping mall, users typically focus on comparing dishes from each restaurant and reading reviews from other customers without the need for navigation, which implies they do not need to switch to other applications. Conversely, when driving in a vehicle, users are likely to start navigation after selecting a desired restaurant. Many food service applications on smartphones do not have built-in navigation functions; therefore, users must switch to other dedicated navigation software, which disrupts the user experience flow. By contrast, in some automotive HMI systems, map navigation and restaurant searching are integrated into the same application, thereby eliminating the need for users to switch between applications and providing a more seamless user experience.

Although services in automotive scenarios are faced with challenges such as driver distraction or cross-application demands, designing automotive intelligent scenarios also has an advantage that the user's intention can be determined more efficiently. When a user picks up their smartphone and unlocks the screen at home, they may want to use a video application for entertainment, contact friends through WeChat, check the weather, or perform one of many other functions. Accurately determining the user's intention is difficult for a smartphone. However, in automotive scenarios, inferring the user's intention is considerably easier. When a user enters the car, they are likely to set a navigation destination. When encountering traffic congestion, they may be interested in some soothing music. When approaching a shopping mall, they may want to learn about special offers by the stores in the mall. By integrating various data and performing computations and predictions, automotive HMI systems can potentially provide users with more proactive, seamless, and intuitive interaction experiences.

1. **Definition of Scenario Storylines**

The series of situations and corresponding behaviors that users encounter while planning a trip and driving or riding in a vehicle until they reach their destination constitute the travel scenario storyline. The most common travel purposes for Chinese automotive users are daily commuting, shopping at malls, urban recreation, suburban outings, and long-distance road trips. In specific travel scenarios, users will have varying needs that arise from factors such as changes in the stage of vehicle usage, driving routes, road conditions, and weather conditions, among others. The scenario

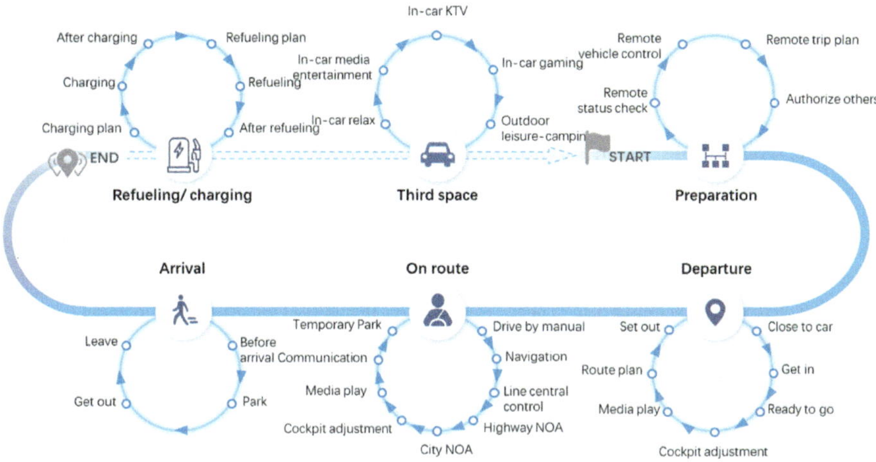

Fig. 9.3 Six stages of a complete travel scenario storyline

storyline generally includes the following stages of vehicle usage: preparation, departure, on route, arrival, refueling or charging, and third space, as shown in Fig. 9.3. However, some storylines may contain only three to five of the aforementioned stages.

Next, we will introduce a typical scenario storyline for a daily commute. The protagonist is Nick, who lives in a city and drives an electric car to work on weekdays.

At 7 o'clock on a winter Monday morning, Nick wakes up at home. In the preparation stage, he first turns on his cell phone to check the weather and road conditions. The Monday morning rush hour traffic is as expected. Nick estimates his departure time while quickly freshening up and getting dressed. At this point, if the car can proactively send traffic information to Nick's phone for his commute to work, he will only need to tap the screen to learn about this information, without having to locate and activate the navigation application on his phone and enter the address.

In the departure stage, Nick heads to the underground garage, finds his car in the designated parking spot, unlocks it with the key, and enters the car. After entering the car, he places his laptop bag and cell phone in their designated places. After an entire night in the cold garage, the steering wheel and leather seats have become extremely cold. Nick first turns on the climate control system to warm up the vehicle, which also makes his hands more flexible for gripping the steering wheel. In this case, if the vehicle supported remote climate control, Nick could have turned it on 10 min before leaving his house to ensure a warm and comfortable environment upon entering the car. After adapting to the temperature inside the car, he leans forward to activate the in-vehicle navigation software, enters his work address, and selects the most time-efficient route for navigation. Subsequently, Nick opens the in-vehicle music app, finds his favorite playlist, and begins to play it. Once everything is ready, he shifts the gear to Drive, checks his surroundings, and drives out of the parking space.

During the on route stage, Nick needs to pass through residential roads, urban roads, and expressways. When he exits the community, he notices that the visibility is poor. Therefore, he pulls over by the road near the community gate, opens the in-vehicle weather application to check the air quality index, and finds a yellow warning for haze. Thus, he turns on the air purification function of the car's climate control system before proceeding onto the main urban road. If the vehicle could proactively remind the driver to activate the purification function when it detects poor external air quality, Nick could have avoided the need to stop the car to perform complicated operations.

After Nick drives on urban roads for a while, the road becomes congested because of the morning rush-hour traffic, with an endless line of red lights ahead. He wants to take advantage of this time to pre-order breakfast from his favorite coffee shop on the bottom floor of his office building such that he could pick it up directly after arriving. Ordering food on a cell phone is cumbersome, and he has to occasionally pay attention to the distance from the vehicle in front, which causes Nick to feel somewhat anxious. At this point, if the vehicle was equipped with a food service application that can display several recommended food options based on the user's order history, Nick could simply glance at them, select a few desired items, and complete payment using the car's voice interaction system, thus avoiding excessive visual distraction and maintaining situational awareness. Once the breakfast has been successfully ordered, the road also becomes clear. With the music playing, Nick drives onto the urban expressway. He is very familiar with this road; therefore, he does not need to constantly check the navigation information. After driving for more than 10 min, he reaches the expressway exit and finds that the usually smooth exit is now crowded because a vehicle ahead had been scratched when changing lanes. Nick regrets not taking the previous exit, as waiting for two additional traffic lights would have been faster than being stuck in the current queue. Nick then increases his speed to avoid being late for work. In fact, the car could have proactively alerted Nick prior to the previous exit that the road ahead was congested and advised him to leave the expressway earlier to reduce the waiting time and arrive at work faster.

After a journey of more than 30 min, Nick finally arrives at the company parking area and enters the parking stage. At this point, the vehicle has only 30% of the battery remaining and Nick hopes to fully charge it at his workplace before driving back home later. He drives to the charging piles in the parking area, only to find that all of them are occupied. Thus, he has no choice but to park temporarily in another parking lot and plans to check for an opportunity to charge his car at noon. If the vehicle had proactively inquired about the need for charging when the battery level fell below a certain threshold and provided options to reserve a charging pile, Nick would not have to worry about not having time to move his car to the charging pile because of an unscheduled meeting at noon, thus avoiding anxiety.

After parking the car, Nick unbuckles his seatbelt, takes his laptop bag and phone, and opens the door to step out. Nick's phone suddenly rings and it turns out to be a colleague asking about the location of the meeting room he had reserved. He tries to recall the room number and briefly chats with his colleague about the meeting content while thinking about the route to the coffee shop. When Nick arrives at the coffee shop and ends the call, he suddenly realizes that he may have forgotten to lock the car, so he has to go back and check. If he could view and control the car's door lock through a mobile application, he would not have to spend time returning to the parking area.

Similar scenario storylines can be designed for various situations, such as Picking up kids from school, meeting with friends on weekends, or going on a family outing to the suburbs. In these storylines, our primary focus is not to evaluate individual car technologies and functions but to assess whether these functions can seamlessly integrate with the previous and subsequent tasks in the scenario storyline and provide a more efficient, smooth, and seamless experience.

2. **Number of Operation Steps and Intelligence Level**

Various criteria can be applied to assess the intelligence level of automotive HMI systems, such as the total time required by users to complete a set of tasks or the degree of satisfaction they experience after completing a task. Under current technological conditions, the number of operation steps is an index that is highly correlated with intelligence level. Fewer operation steps needed to complete a task indicate a higher intelligence level. In human-to-human communication, we cannot necessarily equate lower communication content with higher intelligence when achieving the same communication purpose. This is because effective interpersonal communication often involves not only improving efficiency but also demonstrating proper etiquette, expressing emotions, and fostering mutual empathy. However, as the intelligence level of vehicles is still far below that of humans, it is not necessary to impose such high requirements. If vehicles can achieve optimal efficiency, they will be able to meet the majority of user requirements under different scenarios. The most intuitive performance index of operational efficiency is the number of operation steps. Therefore, if we can only use a single objective and quantitative index to describe the intelligence level of an automotive HMI system in achieving specified goals, the choice should be the number of operation steps.

Optimizing the number of operation steps through intelligent means can be achieved in two ways. The first approach is proactive recommendations. For example, during mealtime, if the map can automatically display a gateway for restaurant selection, users can directly tap on it to search for nearby restaurants, as shown in Fig. 9.4. Otherwise, users would need to enter the directory to search for points of interest and select a restaurant, which requires 2–3 additional steps. If the system can recommend a specific restaurant based on the user's preferences, it can also save them the trouble of searching through a list. However, such a design requires a higher intelligence level to achieve precise recommendations, as inaccurate recommendations may confuse the users. The second approach is using scenario-based modular designs. For example, when users want to take a nap inside their vehicle, they typically need to close the sunshade, control the windows, set the alarm, and adjust the music and lighting, among other tasks. In a traditional interaction logic tree, these functions are distributed in different locations, and some functions are located at deep logic levels, requiring users to perform several tedious operations. By contrast, if these functions are integrated into a "nap mode" widget or shortcut directory, users can simply activate this mode to conveniently operate all functions, significantly reducing the total number of operation steps, as shown in Fig. 9.5. Furthermore, if the system can automatically detect when the user wants to rest inside the vehicle after parking, it can automatically enter the nap mode, further reducing the operation steps and enhancing the intelligence level.

In the two cases mentioned above, we can observe that, under current technological conditions, good contextual intelligence can be simply achieved by using excellent design and simple logic. However, for contextual intelligence to achieve its fullest potential, we must rely on richer data inputs and more powerful intelligent algorithms.

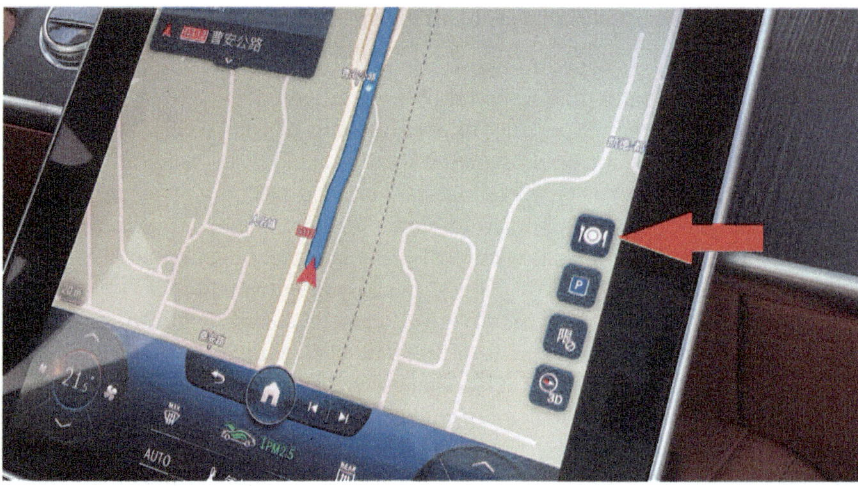

Fig. 9.4 Restaurant selection gateway in Mercedes-Benz S-Class (2020), which automatically pops up at mealtime

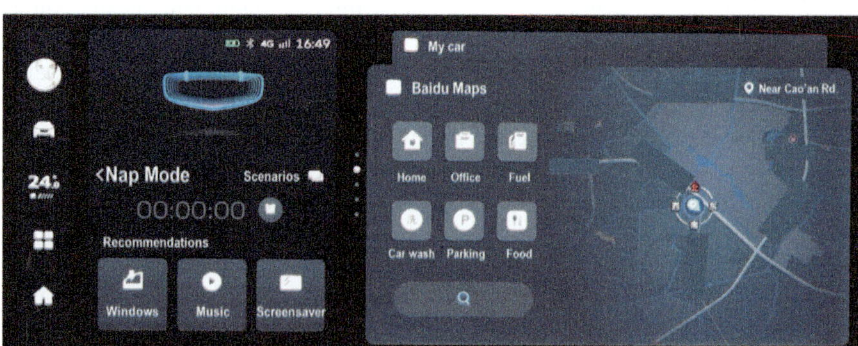

Fig. 9.5 Nap mode widget in the 2021 Geely Xingyue L, translated from Chinese language

When counting the number of operation steps, in-depth research on how to define "one step" should be conducted. We can consider a single tap on the central information display, a press of a physical button, or a brief voice control command as one operation step. However, performing precise sliding gestures on the central information display, such as increasing the climate temperature by 5° in some car models, is more challenging than a single taps. Similarly, stating a specific navigation address is more complicated than issuing quick commands such as "confirm" or "cancel." Therefore, for more complex interaction steps, we can consider assigning them a coefficient greater than 1 to calibrate the actual operational load of such steps.

In automotive HMI systems, certain purely entertainment-oriented functions, such as video games and casual voice chats, do not necessarily need to pursue efficiency. Therefore, the intelligence level of these functions cannot be assessed through step counting and usually requires a more subjective and non-standardized evaluation method. However, these entertainment-oriented functions are not essential in automotive HMI systems, and, even if they exist, do not constitute a significant proportion.

3. **Other Indexes of Contextual Intelligence**

In addition to the two indexes directly related to the number of operation steps, namely, proactive recommendations and scenario-based modular design, contextual intelligence also needs to consider the sense of immersion, personalization, and privacy protection. A sense of immersion refers to the comprehensive atmosphere created by the system's functions or services in a specified scenario, enabling users to immerse themselves in an enjoyable experience. Examples include extensive ambient lighting, captivating on-screen visuals, and surround sound effects. Personalization means that, in a specified scenario, the system can provide targeted and differentiated functions or services for different users at different times and in different environments, such as automatically adjusting the seat position according to the user's body size or recommending familiar restaurants and dishes when entering a commercial district. Privacy protection refers to the security functions or services provided by the system to protect the user's privacy. On one hand, this includes compliance with corresponding data security regulations. For example, cars sold in China should ensure that the images from the car's external camera cannot be directly transmitted outside the vehicle (e.g., to the cloud or the user's cell phone), only if human faces and car number plates outside are blurred. On the other hand, it also involves providing users with a subjective sense of privacy, such as blocking in-vehicle cameras with physical covers.

References

1. Wechsler D. The Measurement of Adult Intelligence [M]. Williams & Wilkins Co, 1939.
2. Erika Hall. Conversational Design [M]. Tsinghua University Press, 2019.

Chapter 10
Values

10.1 Culture and Values

Different groups of people may have different demands for the same type of product. This phenomenon may not have been as noticeable for automotive products manufactured before the era of intelligence. Any given user would want a vehicle with large space, fast acceleration, low noise, and good ride comfort. The performance of these attributes is positively correlated with the vehicle's price. However, in the era of intelligence, vehicle experiences can lead to even greater differences, further highlighting consumers' varied demands for automotive products. For instance, some users might prefer larger screens and fewer physical buttons, while others may wish to retain as many physical buttons as possible; some users love cool animations and lighting, while others desire a simple and clear-cut screen interface. These differences are not related to price and are even minimally related to usage scenarios—they are almost purely personal, subjective preferences. For different groups with limited communication, such as users from two different countries, these personal subjective preferences are even more pronounced. In short, Chinese users may like a particular automotive HMI design that German users dislike, and vice versa.

Intelligence has made differences in the preferences of car users more substantial and important, yet the automotive industry lacks the research experience on these purely subjective preferences. Therefore, we need to explore more comprehensive, systematic, and forward-thinking research approaches.

10.1.1 Cultural Influence on Automotive User Experience Design

When studying the differences in user preferences, the automotive industry typically employs user surveys to obtain direct feedback. However, such research methods have

J. Ma and Z. Gong, *Automotive Human-Machine Interaction (HMI) Evaluation Method*, Research on Automotive Intelligent Cockpit,
https://doi.org/10.1007/978-981-99-9951-4_10

their limitations and are not ideal for the study of subjective preferences, especially for the digital experience. First, user surveys tend to capture superficial opinions rather than in-depth explanations. For example, users can respond whether they need to sing karaoke with friends and family in the car, but they might not be able to articulate why karaoke is more suitable for entertainment with their loved ones than watching movies. Second, users can only express their current thoughts but not predict their future thoughts. This is partly because users have no need or obligation to fully imagine the future, and their future choices might be influenced by other factors. For instance, before the launch of the iPhone X in 2017, users would not have expressed a preference for the "notched display" design, yet this design later became very common in the market. Finally, user opinions are discrete and lack a systematic framework. We can use statistical methods to analyze trends in user group choices, but the interpretation of the underlying causes often relies on the researcher's subjective experience.

Therefore, it is necessary to determine what approach should be adopted to investigate the subjective preferences of users for automotive intelligence experiences. As these are personal preferences, which are essentially differences among individuals, the research should begin by exploring the root cause of these individual differences—culture. The Dutch sociologist Geert Hofstede pointed out that culture is the collective programming or "software of the mind" that distinguishes the members of one group from the others [1].

The uniqueness of people's software of the minds can be divided into three levels: human nature, culture, and individual, as shown in Fig. 10.1.

Human nature is universal and inherited genetically. Most human beings agree that happiness is better than sadness, health is better than disease, abundance is better than scarcity, freedom is better than oppression, and knowledge is better than ignorance. Everyone enjoys listening to stories, myths, and proverbs; all children are fearful of the unknown and unfamiliar; and all adults are more inclined to trust the

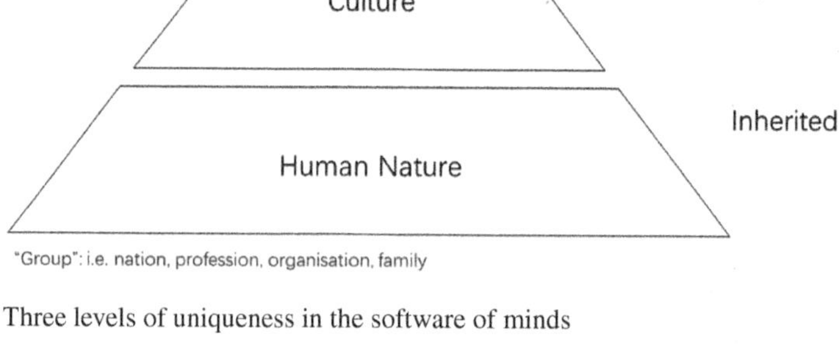

Fig. 10.1 Three levels of uniqueness in the software of minds

members of their group than those outside it. Besides, Immanuel Kant argued that, although the differences in intelligence between individuals cannot be eliminated, there are universal and common aspects in their mental world, that is, esthetics. Thus, the pursuit of beauty is universal to all human beings.

Culture affects all our thoughts, feelings, and actions, which not only includes activities to refine our minds but also mundane matters such as greeting people, eating, and expressing emotions. Culture is a collective phenomenon because it is at least partly shared with people who live or lived within the same social environment, and this environment is where culture is learned. Hofstede believes that culture is learned, not innate. It is derived from an individual's social environment rather than from their genes [1].

Individual is a unique set of mental programs that are not shared with any other individual. It is based on the traits that are partly inherited within the individual's unique set of genes and partly learned.

When studying the subjective and differentiated preferences of users for automotive intelligence experiences, culture serves as the most crucial starting point. However, human nature is shared by all human beings and does not directly result in differentiated user needs, while individual is unique to each person and can vary significantly, even within a small group. However, it is impractical to research every individual separately when defining consumer products. Furthermore, prioritizing the analysis of individual differences over the commonality of a particular group based on culture is not only unnecessary but will also come at the expense of our ability to generalize and predict common patterns.

10.1.2 Values as the Core of Culture

Next, we will delve into the concept of culture. What, precisely, is culture? Collectivism can be regarded as a prominent aspect of Chinese culture, so can the tradition of tea-drinking. Similarly, brainstorming can be considered a part of work culture, so can the conservation of paper. However, it is evident that the key concepts in these instances are not on the same level. In fact, culture can be categorized into four levels: symbols, heroes, rituals, and values, as illustrated in Fig. 10.2 [2]. Symbols represent the most superficial layer, whereas values constitute the deepest manifestation, with heroes and rituals lying in between.

Symbols are words, gestures, pictures, or objects that carry a particular meaning that is recognized as such only by those who share the same culture. The words in a language or jargon belong to this category, as do dressing styles, hairstyles, flags, and status symbols. Several symbols are transient, such as popular words in a language. New symbols are easily formed, while old symbols can disappear.

Heroes are people, alive or dead, real or imaginary, who possess characteristics that are highly prized in a culture and, hence, serve as models for behavior. Benjamin Franklin and Batman in the United States or Lu Xun and the Monkey King in China are all considered heroes within their respective cultures. Heroes are more stable than

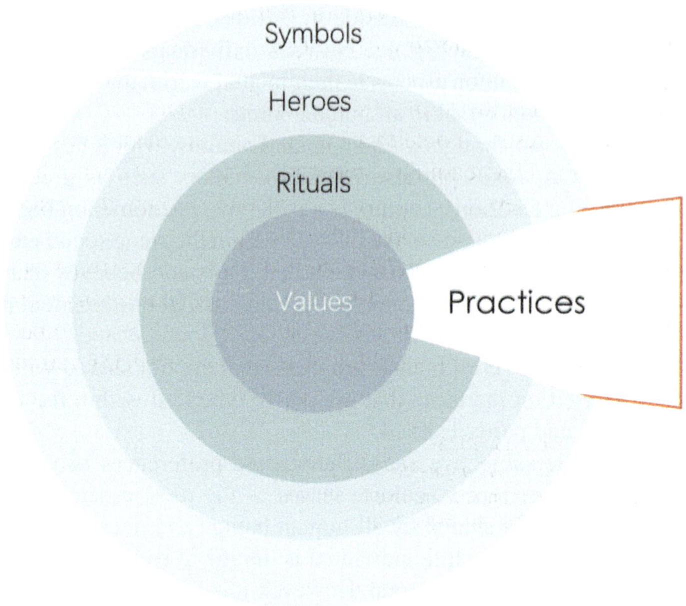

Fig. 10.2 Manifestations of culture at different depth levels (*Source* Hofstede Insights)

symbols. A hero possessing enduring vitality may continuously generate or give rise to new cultural symbols.

Rituals are collective activities that are technically superfluous to achieve desired goals but are considered socially essential from a cultural perspective. Therefore, they are performed for their own sake. Examples include ways of greeting and paying respect to others, social and religious ceremonies, and meetings intended to reinforce group cohesion.

In Fig. 10.2, symbols, heroes, and rituals are subsumed under the term "practices" because they are visible to an outside observer; however, their cultural meaning is invisible and can only be understood in the way these practices are interpreted by the insiders.

A culture's core is formed by values. Values are broad tendencies that prefer certain states of affairs over others. Values are feelings with an added directionality indicating a positive and a negative side. They deal with pairings such as evil versus good, dangerous versus safe, forbidden versus permitted, indecent versus decent, immoral versus moral, unnatural versus natural, abnormal versus normal, paradoxical versus logical, and irrational versus rational.

Any phenomenon related to culture necessarily encompasses both the visible layer of practices and the unseen core of values. For instance, when Chinese people gift several boxes of milk to relatives during the Chinese New Year, the underlying value is the reinforcement of familial relationships within the context of collectivism. The ritual is the manifestation of reciprocation; the choice of gifting milk may be influenced by the spokesperson (i.e., the hero), and the red packaging symbolizes the New Year's festivity. However, the three layers of practices–symbols, heroes, and rituals do not necessarily appear simultaneously in every specific cultural phenomenon. Rituals are usually only related to dynamic processes instead of static states; for instance, symmetric architectural layouts have little to do with rituals. Heroes play a significant role in culture, often triggering the abrupt change of cultural symbols, yet they are not essential to every cultural phenomenon, particularly for more stable cultural phenomenon. For example, when choosing to wear red clothes for the New Year's celebration in China, there is no need to associate this choice with a famous figure from ancient times.

10.1.3 Hofstede's 6D Cultural Model

Values constitute the core of culture. There are numerous approaches to the study of values, among which Hofstede's 6D cultural model is particularly well known. The Dutch sociologist Geert Hofstede, a professor at Maastricht University in the Netherlands and a former researcher at IBM, first proposed his cultural theory in 1980, which has since evolved into the Hofstede 6D cultural model we use today. Hofstede's cultural research results are cited approximately 13,000 times per year in academia, with more than 240,000 citations to date, making him the most-cited social scientist across Europe.

Hofstede's 6D cultural model can abstract the cultural characteristics of any country as scores across six dimensions. Hofstede transformed the field of culture, once a predominantly qualitative research area, into a quantitative research area, making it more easily comprehensible, applicable, and comparable as well as significantly lowering the threshold for cultural research within the field of sociology. This theory has been applied across various fields such as human resource management, international trade, and experience design.

The six dimensions in Hofstede's model are power distance, individualism versus collectivism, masculinity versus femininity, uncertainty avoidance, long- and short-term orientation, and indulgence versus restraint [2]. Each dimension is scored as an integer between 0 and 100, calculated through quantitative surveys. The scores in Hofstede's 6D cultural model cover more than 70 countries and regions worldwide. Specific figures can be found on the official website of Hofstede Insights (https://www.hofstede-insights.com/). It is worth noting that there is no right or wrong judgment in high or low scores on any dimension. These are merely relative values representing the differences in cultural values.

Power distance refers to the extent to which less powerful members of an organization accept that power is distributed unequally. It reflects people's attitudes towards inequality. Cultures with high power distance tend to see hierarchies as necessary, accept the privileges that come with power, obey their superiors' instructions, and agree with centralized decision-making. Conversely, cultures with a low power distance are inclined to view hierarchies simply as convenient, do not approve of privilege for anyone, exhibit stronger initiative, and support decentralized decision-making.

Individualism refers to societies in which the ties between individuals are loose and everyone is expected to look after themselves or their immediate family members. By contrast, in collectivist societies, people are integrated into strong, cohesive in-groups that look after them in exchange for loyalty. In the study of societal culture, a significant negative correlation can be found between individualism and collectivism; hence, the two factors can be considered a single dimension. Individualism and collectivism embody the extent to which people depend on others. Individualist cultures tend to prioritize tasks over relationships, show loyalty to close family members, communicate explicitly, value freedom, and feel guilt over mistakes. Conversely, collectivist cultures are inclined to prioritize relationships over tasks, show loyalty to their collectives, communicate implicitly, strive for harmony, and feel shame over mistakes. Countries with individualistic cultures are more likely to have a low power distance, with some exceptions, such as France and Belgium, which are individualism societies with relatively high power distance.

Masculinity cultures prioritize success and achievement, whereas femininity cultures focus on interpersonal relationships and quality of life. Masculinity and femininity embody the sources of people's motivations. This dimension is less intuitive and difficult to understand, yet remains crucial. In masculinity cultures, life is often perceived primarily in the context of work, success is admired, dominant views are expressed, ambitious goals are pursued, and a result-oriented mentality is emphasized. Conversely, in femininity cultures, work is considered as a means to enhance the quality of life, sympathy is shown towards the unfortunate, consensus and cooperation with others are valued, the goal is to pursue a higher quality of life, and a process-oriented mentality is emphasized. It is worth noting that, although the terms "masculinity" and "femininity" have gender connotations, the value orientations they represent are not inherently related to sex and certainly should not be equated with male or female chauvinism. For example, a man who is good at coordinating multiple interests and achieving work–life balance fits the description of femininity, but this does not necessarily mean he is a "feminine" man. Furthermore, masculinity is occasionally confused with high power distance, but the two are significantly different. High power distance means that the weak accept the gap with the strong, whereas masculinity represents the desire to become strong. The gap in power distance between individuals is often inherent and difficult to reduce, whereas the gap in masculinity between individuals can be narrowed through personal effort.

Uncertainty avoidance is the extent to which people feel threatened by ambiguous or unknown situations. This dimension reflects how people deal with the unknown. Cultures with high uncertainty avoidance tend to choose traditional things, fear change, be forced to innovate, need authoritative figures, have the self-drive to work hard, and use more deductive logic. By contrast, cultures with low uncertainty avoidance tend to choose trendy things, accept change, innovate spontaneously, do not necessarily need authoritative figures, work hard when needed, and use more inductive logic. In today's rapidly changing society involving numerous domains such as economics, politics, employment, and consumption, people are continuously creating and consuming new products, which makes the uncertainty avoidance dimension increasingly important. Of note, low uncertainty avoidance does not mean people like or prefer uncertainty. In fact, everyone is afraid of and resistant to uncertainty, but some are more willing to bear higher levels of uncertainty when presented with potential gains. For instance, no one wishes to experience a mobile software failure, but some would choose to update to an unstable new version to try new features as soon as possible. Additionally, uncertainty differs from risk. Economic risks can usually be calculated statistically, whereas uncertainty cannot be measured. The British economist John Maynard Keynes pointed out that the issue with uncertainty is that "there is no scientific basis on which to form any calculable probability whatever. We simply do not know." [3].

Long-term orientation emphasizes nurturing and encouraging virtues that are oriented towards future rewards, particularly perseverance and thriftiness. Short-term orientation encourages the pursuit of virtues related to the past and present. Long-term versus short-term orientation reflects people's attitudes toward time. Cultures with a long-term orientation tend to work for the future, value perseverance, believe in the flexibility and diversity of reasoning (encapsulated in the Chinese saying "specific issues require specific analysis"), and emphasize obligations. Conversely, cultures with a short-term orientation are inclined to work for the present, expect quick results, believe in absolute truths, and emphasize rights.

Indulgence represents the tendency to allow the relatively free gratification of basic and natural human desires related to enjoying life and having fun. Restraint represents the need to suppress and regulate such gratification of desires via strict social norms. Although the term "indulgence" might carry negative connotations in some languages, in this case, it is completely neutral, similar to the other dimensions. Indulgence versus restraint is a measure of how societies manage natural human desires and impulses. Cultures leaning towards indulgence tend to have fewer moral norms, consider leisure time important, spontaneously express emotions, and place more emphasis on outcomes. By contrast, cultures leaning towards restraint often have more moral norms, consider duty and responsibility important, suppress the expression of emotions, and place more value on effort.

Among the six dimensions, the first four appeared in the first edition of Hofstede's "Cultures and Organizations: Software of the Mind", whereas long-term versus short-term orientation and indulgence versus restraint were introduced in the second and third editions, respectively. Therefore, the first four can be regarded as the most original and fundamental dimensions, which already possess strong comprehensiveness and independence. The latter two dimensions can be considered complements that further enhance the comprehensiveness of Hofstede's model. However, these two new dimensions are relatively correlated with some of the earlier dimensions and are not completely independent. For instance, individualistic cultures tend to prefer indulgence, whereas collectivistic cultures favor restraint. Therefore, some scholars have opted to use only the first four or five dimensions when applying Hofstede's model [4, 5]. For example, Huib Wursten's theory of culture clusters categorizes countries worldwide into seven major groups based on combinations of the first four cultural dimensions in Hofstede's model.

Hofstede's 6D cultural model may not necessarily be the only comprehensive or in-depth theory for cultural research, but it is the most user-friendly, particularly for researchers outside the field of social sciences. This is because it is highly systemic and quantitative, allowing complex cultural phenomena to be abstracted into a few standardized numerical values. When using Hofstede's model, two points should be noted. First, the primary aim of these cultural dimensions is to provide a holistic perspective to guide our corresponding analyses and deliberations. The scores are simplifications and abstractions of the analysis results under each perspective. Hence, researchers should not merely focus on comparing the scores while overlooking the specific underlying values and practices. Second, although culture is important, it is one of the influencing factors in the process of social development and cannot replace the impact of other elements such as economics, politics, and technology.

10.2 Typical Values of Intelligent Vehicle Users

Compared with users in other markets, Chinese automotive users have several distinctive demands for automotive experience design, particularly for HMI design. The emergence of these demands can largely be traced back to the core values of Chinese culture. Figure 10.3 shows Hofstede's 6D cultural model scores for China, Germany, Japan, and the United States. It is evident that China has significant differences in most dimensions compared with other countries, and these differences are precisely the fundamental reasons for the different needs of Chinese users.

The six dimensions in Hofstede's model are highly abstract and cannot be used to directly guide or evaluate the user experience design. Therefore, we need to analyze and organize these dimensions in a more concrete manner to obtain the typical values of Chinese automotive users. These values reflect the subjective feelings that users

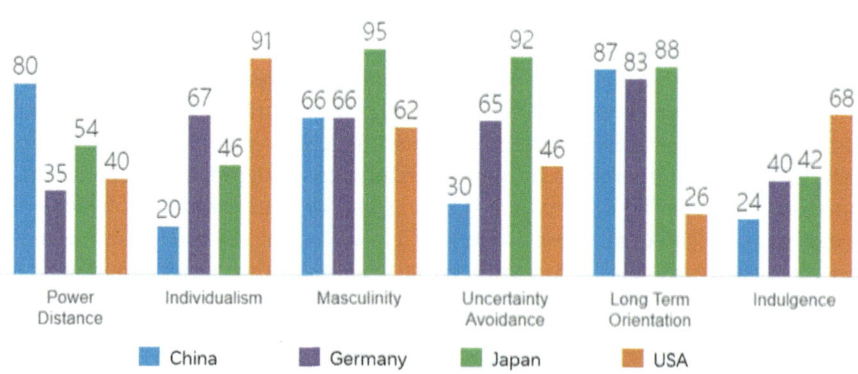

Fig. 10.3 Hofstede's 6D cultural model scores for different countries (*Source* Hofstede Insights)

hope to gain when purchasing and using vehicles. Although these values are not as abstract as the six dimensions in the model, they still belong to the innermost level of cultural manifestation shown in Fig. 10.2. The specific manifestation of these values in products primarily involves the two levels of practice: rituals and symbols.

Compared with European users, Chinese automotive users hold several unique values. In this section, we will mainly introduce the three most typical values: companionship, honor, and surprise and delight. There are two reasons for choosing these three values: first, Chinese users pursue these values more strongly than users in other major markets worldwide, and the reason for this is attributed to the uniqueness of the Chinese culture in Hofstede's cultural model. Second, these values are highly relevant to the automotive product experience, especially the automotive HMI experience. They can be easily realized in automotive products and form differentiated competitive advantages.

The analysis of these three unique values not only facilitates our understanding of the values of Chinese automotive users but also provides ideas for reference when studying automotive users in other countries. This can help product designers analyze the needs of different users in a more objective, comprehensive, and efficient manner.

In addition, reassurance is a value that is crucial to intelligent vehicle development worldwide and will also be discussed in detail in this section.

10.2.1 Companionship

Companionship is a value that is primarily reflected in the automotive HMI as the ability to allow users to interact with other people or anthropomorphic avatars. Companionship primarily stems from China's high collectivism culture or, in other words, its low individualism culture.

Chinese people are more willing to live in larger collectives and interact frequently with the members of these groups. Many Chinese individuals have very close relationships with relatives outside their immediate family. Group entertainment activities, such as square dancing, are very popular in China. Even commercial films expressing heroism often center around a small collective, such as "The Wandering Earth" and "Operation Red Sea," which is completely different from typical Hollywood superhero films. For contemporary young Chinese users, the sense of companionship remains highly significant but is also becoming increasingly complex. They no longer desire an all-encompassing sense of companionship but instead are seeking timely and moderate companionship, which allows them to feel at ease. Additionally, the groups with which these young users identify are shifting from familial relations to interest-based circles.

Companionship is essentially not involved in the basic usage of vehicles. Whether it is driving the car or using the HMI system for navigation and listening to music, these tasks are performed by the driver alone, without the need to involve others or the use of anthropomorphic figures to achieve clearer information transmission. Therefore, companionship is not a necessity in the automotive experience but rather a bonus. However, the importance of companionship should not be overlooked. In fact, certain automotive products that provide users with companionship have achieved remarkable success in the Chinese market and are now popular case studies.

The most common manifestation of companionship is an anthropomorphic (or zoomorphic) voice assistant avatar. Compared with other interaction modalities, voice control in automotive HMI is more suited to the application of avatars. First, the voice control interaction process is more similar to a conversation between people, in which the introduction of an interactive avatar can make users feel more natural. By contrast, the interaction process of touch screens and buttons involves step-by-step commands, which are somewhat different from human conversation. Furthermore, voice control does not have a visible interface, making it difficult for users to perceive the interaction status of the system. A voice assistant avatar can display the interaction status, such as listening, processing, speaking, and so on.

Voice assistant avatars can be divided into anthropomorphic (or zoomorphic) and abstract types. Anthropomorphic voice assistants will present a human or animal face to display emotions, lip shapes, and other features, with some even having a body and limbs to show body language and personalized attire. Currently, several mainstream Chinese car brands have adopted anthropomorphic voice assistants. In fact, anthropomorphic or zoomorphic mascots are widely used by Chinese companies in various fields. For instance, Tmall has a cat-like mascot and JD.com has a dog-like mascot, whereas international mainstream shopping websites such as eBay and Amazon do not have similar mascots. Abstract voice assistants are typically dynamic geometric shapes or waves used to display the interaction state. A typical abstract assistant is Apple's Siri, while many European and American car brands also use abstract voice assistants, as shown in Fig. 10.4. An avatar similar to a human or animal feels very similar to having a family member, a companion, or a pet communicating and interacting with the user, thus generating a stronger sense of companionship. Conversely, an abstract voice assistant avatar primarily displays the interaction state, making it more difficult for users to feel a sense of companionship.

Anthropomorphic (or zoomorphic) voice assistants can be further classified. From an image realism perspective, they can be classified according to style as sketches (e.g., Li Auto One shown in Fig. 10.5), cartoons (e.g., Xpeng P7 shown in Fig. 10.5), and full simulations. Theoretically speaking, the more realistic the avatar, the stronger the sense of companionship it conveys; however, it is also vital to avoid the "uncanny

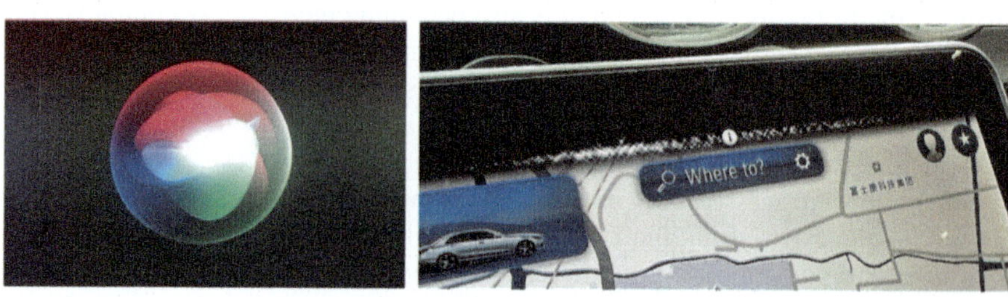

Fig. 10.4 Voice assistant avatars of Apple's Siri and Mercedes-Benz C-Class (2023)

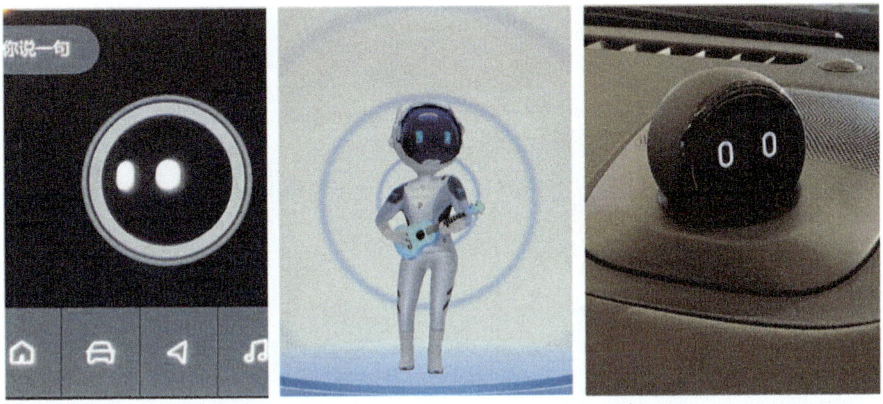

Fig. 10.5 Voice assistant avatars of Li Auto One (left), Xpeng P5 (center), and NIO ET7 (2021) (right)

valley" effect. In terms of image completeness, they can be primarily classified as eyes (e.g., Li Auto One shown in Fig. 10.5), mainly eyes and mouth, a complete head, and a complete body with limbs (e.g., Xpeng P7 shown in Fig. 10.5). The more complete the avatar, the more human-like it looks. However, the eyes are the easiest way for the voice assistant to convey rich emotions, and complex mouth shapes and body language might distract the user. Regarding the display format, the voice avatar can be displayed on a screen, as a small physical robot (e.g., NIO ET7 shown in Fig. 10.5) or as a holographic projection. Small robots can easily attract users' attention and become a prominent selling point for the product; however, owing to hardware constraints, there is limited room for innovation. By contrast, on-screen or holographic avatars can be more creative in their dynamic expressions and accessories. For example, the avatar Vicky designed by AMMI for the Volkswagen:UX prototype cockpit can appear on the screen or hover above the dashboard and has approximately 30 different expressions and accessories, as shown in Fig. 10.6.

In addition to having anthropomorphic avatars, voice assistants can also interact with users for recreational purposes, such as chatting, telling jokes, playing riddles, and other activities. These features can help users feel a sense of companionship similar to that of real humans. The current usage frequency of these features is relatively low, partly because several users are not accustomed to chatting with machines and partly because the dialogue capability of voice assistants is not as natural as that of real humans. With the cultivation of user habits and the advancement of technology, the importance of these features will gradually increase.

Apart from interacting with voice assistants, another important direction in companionship is to facilitate user interaction with other people. For example, if the HMI system supports mainstream social messaging software, it not only allows having conversations with friends but also enables the sharing of driving-related information such as the current location, destinations, and driving routes. Driver distraction is a limiting factor for the widespread use of in-vehicle social software.

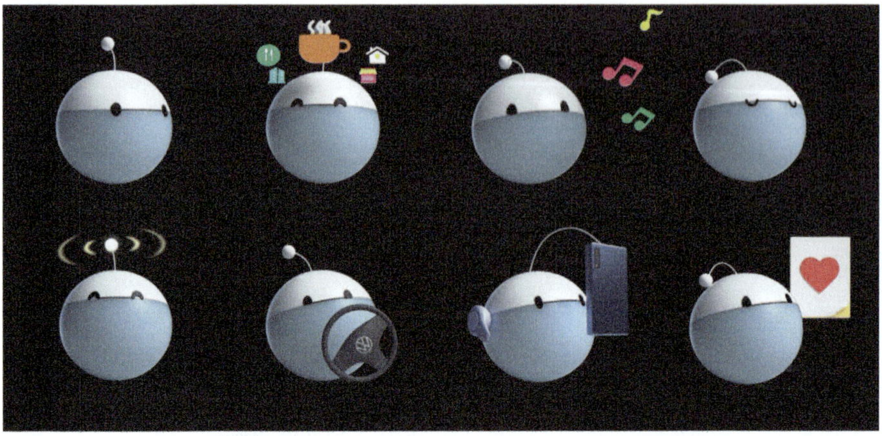

Fig. 10.6 Wide variety of expressions and accessories displayed by the voice assistant of the Volkswagen:UX prototype cockpit (2021) (*Source* Volkswagen Group China)

Currently, whether it is in-vehicle WeChat or the SMS feature in Apple CarPlay, these can only support the voice input and output of conversational content, which greatly reduces visual distractions, but also significantly lowers conversational efficiency, causing the reluctance of some users to use these applications.

Phone calls are another means of connecting in-vehicle users with other people. This method is simple and direct; however, several automotive HMI systems have overlooked its importance. Adding a feature on the home page or in the shortcut menu to call designated contacts with one tap would make users who are accustomed to socializing on the phone feel that contacting their friends and family is right at their fingertips.

Interaction among in-vehicle passengers is also important. Many car models are equipped with an increasing number of screens, not only for the driver but also for the front and rear passengers. Each passenger's screen can provide individual audio–visual entertainment while also serving as a tool for interaction among vehicle occupants to enhance task efficiency and foster emotional bonds. For instance, the front passenger could browse restaurant information on their screen to make a selection and send the chosen address to the navigation system, which in turn would guide the driver to the selected destination, as shown in Fig. 10.7. This not only avoids driver distraction caused by searching for restaurants but also increases the interaction among passengers. Passenger interaction between multiple screens not only requires product designers to fully investigate and define various scenarios but also places higher demands on the hardware architecture of the HMI system.

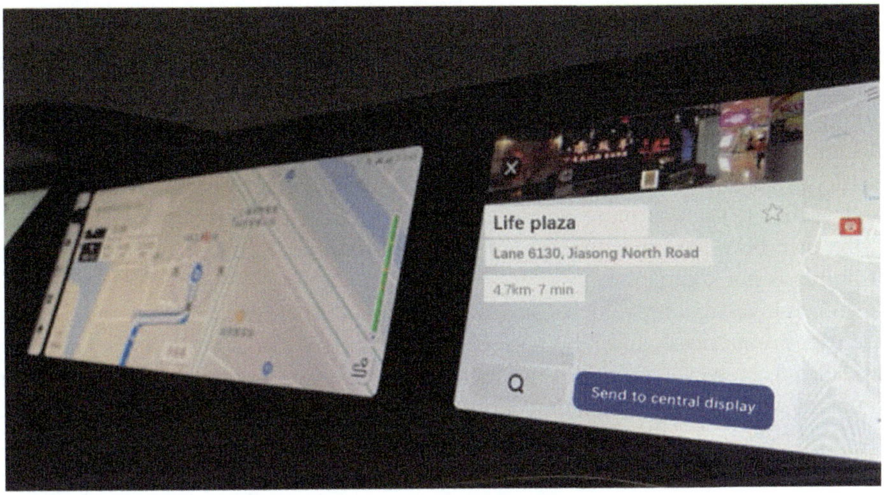

Fig. 10.7 Ability to send destinations searched from the front passenger screen to the navigation system on the central information display in Geely Xingue L (2021), translated from Chinese language

10.2.2 Honor

In an automotive HMI, honor as a value is primarily embodied by the use of eye-catching or achievement-symbolizing designs to enhance the user's confidence and sense of pride as well as fostering the respect and recognition of others. This value primarily arises from the high masculinity and high power distance of Chinese culture. The former dictates people's motivation to pursue honor and the latter explains why products have sufficient diversity to reflect honor.

Concerning masculinity, Chinese people generally agree that everyone should have dreams, work tirelessly, overcome their competitors, and achieve higher accomplishments. Therefore, long overtime hours are common in Chinese companies, and successful entrepreneurs and founders are idolized. The pursuit of achievements naturally implies that people are more willing to showcase their accomplishments, such as by purchasing luxury goods to show off. The ostentation engendered by the masculine culture is focused on the price of the item rather than the enjoyment of its functionality. For example, some people buy Rolex watches not because they like its specific style but because its price and the wealth it symbolizes are widely recognized.

Understanding the manifestation of honor in the automotive user experience also requires studying the power distance culture. Chinese people are subjected to ranking by exam scores in schools and encounter various performance assessments at work. Even the number of steps walked each day can appear on a leaderboard in WeChat. Within an organization, the leader not only manages the team but often also has a high level of authority and enjoys more resources such as a larger office area. Therefore, in the design of products and services, there will be a rich variety of differentiation to reflect different levels or grades. For example, the German researcher Martin Karaffa found that, in countries with high power distance, premium car brands have a higher market share. These brands, including Mercedes-Benz, BMW, and others, do not exhibit the same consumption characteristics as watches, bags, and other luxury goods. The high price of premium cars usually includes better functionality and usability, such as more power and more comfort configurations, whereas the difference between other luxury and ordinary goods in terms of functionality and usability is not significant.

When investigating vehicles as a type of consumer product, we can observe that the user preferences caused by masculinity and power distance are indeed strongly associated. However, this does not mean that these two dimensions are similar. In domains where other cultural models are applied, such as corporate management and business communication, these two dimensions have very significant differences, and the similarity manifested in the field of consumer goods is just a special case.

Honor is a subjective perception and does not intrinsically embody any functionality or usability attributes. However, a specific product design that embodies honor often encompasses functionality or usability as well. For instance, the visual impact produced by a larger in-vehicle screen can be associated with honor, while its ability to display a more extensive range of information contributes to its functionality. When analyzing such designs, it is essential to isolate the attributes of honor for a more targeted discussion.

Honor plays a crucial role for many Chinese consumers when choosing automotive products and could even potentially serve as a decisive factor in whether to purchase a specific car model. Larger body sizes are very popular, despite the fact that the rear seats of some vehicles are rarely occupied. Larger wheel hubs are equally appreciated, even though they might amplify the roughness of daily roads. Panoramic sunroofs are also very well-liked, even if the vehicle owner does not particularly enjoy basking in the sun. These features and equipment do not necessarily enhance functionality or usability and may even impede these aspects. However, these aspects are all readily visible to others and, hence, are easy to show off. By contrast, some design elements that enhance personal user experience are not valued by Chinese users but are prevalent in the German market. These include seat ventilation and heating, custom interior leather stitching, and so on.

In automotive HMI, honor can be embodied by hardware devices with sensory impact, the most common among which is to have more and larger in-vehicle screens. Currently, there is no consensus on whether more and larger screens will improve usability. Although these screens can display more information and reduce operational steps, they often increase the cognitive load and the difficulty of touching the screen with one's fingers. Moreover, when the in-vehicle screen area increases exponentially, it becomes necessary to assess what information can and needs to be displayed. These issues are still in the preliminary exploration phase of automotive HMI design, and a consensus has not yet been reached. Therefore, some car models have remained restrained with respect to the size and number of screens, using only two central information displays and an instrument cluster display not exceeding

12.3 inches. Some vehicles even use an instrument cluster display with an area of only approximately 5 inches, even smaller than the screen of current mainstream small-sized smartphones.

However, from the perspective of honor, having several and larger screens offers a clear advantage. The screen serves as the core input and output hardware for the automotive HMI system, naturally becoming an important symbol for a sense of technology in the cabin. Most Chinese users agree that having larger and more screens will convey a stronger sense of technology, and the vehicle will be perceived as more advanced and valuable. This impression is subjective and not necessarily linked to the actual performance and usability of the HMI system. This impression is also critical because when choosing a vehicle, many consumers only spend a few seconds or minutes experiencing a few functions in the HMI system, and such experiences are inevitably subjective, general, and incomprehensive. Therefore, several and larger screens are likely to persist as a prominent trend in automotive HMI development. This is not because users genuinely need to see a considerable amount of information on the screen but because users appreciate the atmosphere and feelings created by the screens, as exemplified by IM L7 shown in Fig. 10.8.

In addition to screens, other HMI hardware can also provide a sensory impact that instills a sense of honor in users. Examples include large-area dynamic ambient lighting, exquisite retractable speakers, buttons and knobs with a crystalline texture, and steering wheels with a distinct, non-circular design.

The concept of honor in automotive HMI can also be manifested in the scoring and ranking of user performance. This approach is widespread in mainstream mobile applications in China, such as the step count ranking on WeChat Sports, the hot

Fig. 10.8 Retractable instrument cluster display, central information display, front passenger display, and lower control display in IM L7 (2022) (*Source* IM Motors)

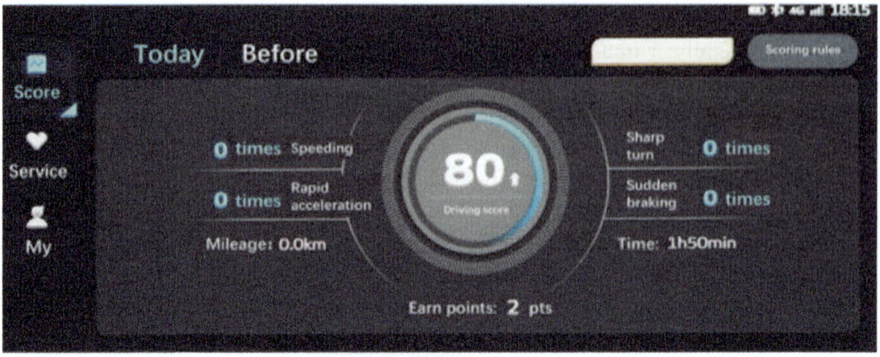

Fig. 10.9 Driving scoring interface of Geely Xingyue L (2021), translated from Chinese language

search list on Weibo, and various gaming rewards and rankings. Some popular navigation apps in China, such as Amap and Baidu Maps, provide rich data and beautifully animated summaries at the end of the navigation, a feature absent in Western counterparts such as Google Maps and Waze. Automotive HMI can offer performance scoring and ranking, as shown in Fig. 10.9, but such designs are still at an early stage. Automotive companies can use scoring and ranking to reinforce their brand attributes. For example, they can emphasize their environmental friendliness by scoring energy consumption or carbon reduction, highlight the worry-free range of electric vehicles by ranking the number of remote destinations, or demonstrate a vehicle's control performance by providing statistics on longitudinal and lateral acceleration. These scores and rankings should have dedicated interface designs to better achieve the theme-related atmosphere and should also be sharable with friends to further emphasize the sense of honor in a social environment.

Owing to cultural differences, the value needs of car users around the world may differ significantly from those in the Chinese market. For instance, in the Nordic concept of Jantelagen, boasting about one's wealth is considered off-putting, and one should instead present oneself as ordinary, without any sense of superiority. In such cultures, certain elements of Chinese automotive design that express a sense of honor might not be suitable.

10.2.3 Surprise and Delight

In an automotive HMI, the values of surprise and delight are primarily expressed by satisfying people's desire for novelty through unique or continually updated designs. This value mainly stems from the lower uncertainty avoidance of Chinese culture, which implies a greater willingness to accept uncertainty.

Chinese people are generally more willing to embrace uncertainty. In terms of career development, many young people aspire to create or join start-up teams. In the realm of consumption, people are eager to try new cell phone brands, new automotive brands, and new tech products. In Europe, many families tend to stick with the same automotive brand in the long term, even opting for successive generations of the same car model. However, in China, many people wish to experience as many automotive brands as possible. The reason they switch brands is not because of any dissatisfaction with the products they previously purchased and used but merely the desire to try another brand. Furthermore, according to a report by Worldpay, more than 50% of e-commerce payments by Chinese users were made through e-wallets in 2016, whereas the proportion of e-wallet payments was only approximately 20% in Europe and America, where traditional methods such as credit cards still dominated. Whether it is career choices or consumer behavior, new options introduce the potential for uncertainty. However, the novel opportunities and experiences they offer are also more enticing, thus prompting users to forsake certainty. Novelty does not necessarily imply better usability—novelty itself is a type of surprise and delight, and an improved usability is yet another form of surprise and delight.

The lower uncertainty avoidance of Chinese culture also provides excellent opportunities for emerging automotive brands. In 2021, 6 of the top 10 best-selling brands in the Chinese A-segment and above electric vehicle market were new brands, despite having been in the Chinese automotive market for less than 10 years. Upon their initial launch, the products of these new brands not only had potential software faults but also features that were temporarily unavailable. Nevertheless, the introduction of new ideas, products, services, and experiences still attracted a substantial number of customers, thereby supporting their growth.

In automotive HMI systems, surprise and delight encourage more proactive innovations in products, which may entail exploring better interactions at the risk of potential failure or simply altering superficial forms to create a sense of novelty.

Over-the-air (OTA) updates for automotive HMI systems have become widespread in recent years. They allow not only the remote fixing of system faults but also the optimization of the system's logical architecture and the addition of new features, thus serving as an important means of introducing surprise and delight to users. For instance, when the Tesla Model 3 was upgraded to the v11 system, significant modifications were made to the layout and colors of the shortcut icons at the bottom of the page as well as the introduction of new features such as in-vehicle karaoke and exterior light shows. OTA updates impose higher demands on the automotive system architecture. Not only do the relevant software and firmware need to be updated under unified coordination, but an almost 100% success rate needs to be ensured during the update process. Furthermore, OTA updates bring new opportunities and challenges to the automotive business model. Under traditional transaction models, consumers pay a one-time fee to receive all functions at once. However, in the new model, the fees paid by consumers cover not only the currently delivered functions but also expectations for future, yet unseen, features. In addition, additional fees can also be charged for some new features implemented through OTA updates.

Surprises can also be delivered to users through greetings and salutations that appear at a specified time or in specified scenarios. For example, playing celebratory videos for users on special occasions such as the New Year, Christmas, or the user's

Fig. 10.10 Video celebrating Christmas played in the BMW's central information display (2020) (*Source* BMW Group)

birthday (Fig. 10.10) or achieving more precise triggering by incorporating factors such as the vehicle's location and weather conditions. The content of these greetings and salutations can be stored in the vehicle's system in advance and triggered at a specified time or scenario and, hence, do not necessarily have to rely on frequent OTA updates. With the technology of AIGC (Artificial Intelligence Generated Content), the contents can even be generated real-time and to achieve the complete personalization. Typically, the content of these greetings and salutations does not generate actual functionality but consists of interesting designs. Additionally, this content should not contain advertisements or other information that may annoy the user.

10.2.4 Reassurance

Reassurance is not only a value demanded by Chinese users but also critical for the development of intelligent vehicles worldwide.

The values demanded by intelligent vehicle users are often varied and dependent on their respective cultures. Nevertheless, there are certain universal characteristics in the values demanded by users worldwide that are determined at the human nature level (instead of the cultural level) of the human software of mind structure depicted in Fig. 10.1. Theoretically, these universal values are abundant, but because they are shared by all users worldwide, most of them are obvious and need no further explanation. For instance, when executing non-gamified tasks, all users strive for higher efficiency under the principle that fewer steps are always better. When entering personal information, all users seek better privacy protection and believe that the storage and communication technology should always be as secure as possible. Different groups of people may pursue universal values, such as higher efficiency and information security, to varying degrees, but these differences are limited. Thus, these universal values are usually considered less important in research compared with value differences resulting from cultural differences. However, with the rapid advancement of automotive HMI systems, users' understanding and habits often cannot keep up with technological advancements. Hence, some universal values and patterns need to be studied in greater depth.

Avoiding uncertainty in development is a common human tendency. In the discussion on the cultural dimension of uncertainty avoidance, we mentioned that everyone is fearful of and resistant to uncertainty. However, when presented with potential gains, some people are willing to experience greater uncertainty. Predictable, controllable designs can eliminate uncertainty and provide users with a sense of reassurance. This sense is the subjective feeling that the system conveys to users rather than the stability and reliability of the system. For example, if a refrigerator door does not have an exposed handle, the user will assume that the handle is hidden on the edge of the door even without proper signs. However, if the door switch is a pedal, such as in a trash can, users will feel that it has subverted their expectations and subjectively believe that it is inconvenient. Another example is the progress bar shown on many mobile applications when online content is loading. This progress bar does not increase the speed of content loading, and its display may even be an illusion (e.g., it can show 99% loaded even when completely disconnected from the internet). However, it can make users feel that the software is working normally, and the progress appears to be predictable and controllable.

Reassurance is becoming increasingly important for automotive HMI systems. With the rapid development of technology, an increasing number of new features and new interaction modalities have emerged. Several users have no experience in using them and, thus, find it difficult to establish expectations. Therefore, automotive HMI design not only needs to achieve functionality and theoretical efficiency, it also needs to guide users in establishing expectations and make them feel that they are in control of the system.

Larger elements and fewer content clusters in the main interface can convey a sense of reassurance. Such a design not only enhances the visibility of elements and the speed of content retrieval but also ensures that users do not feel overwhelmed by the amount of information, thereby preventing subjective anxiety and resistance. The design of larger elements does not only simply mean enlarging the existing elements on the page but can also involve the creation of a new design style. In addition to featuring larger-sized elements, this style often employs vibrant colors, significant color differences between elements, and neatly arranged clusters. Typical examples include smartphones running the Windows 8.1 operating system and Google's Android Automotive operating systems, as shown in Fig. 10.11. However, for various reasons, both operating systems have not achieved a high market presence.

Voice control can also convey a sense of reassurance by displaying the user's spoken input in real time on the screen. With current technology, voice control in automotive HMI is still less natural than human-to-human communication. Issues

Fig. 10.11 Central information display homepage of Polestar 2 (2019) (left, translated from Chinese language) and the HTC Titan II smartphone (2012) (right; image source: HTC)

such as the misrecognition of proper nouns and incorrect judgment of the starting and ending points of a conversation can occur. Therefore, users need real-time feedback to determine whether there are any issues with the interaction process and whether it is under their control.

In addition, users should be provided with tutorials or tips for voice control. For example, when initiating a voice dialogue, the screen can display a message such as "You can say 'Navigate to xxx'," or it can anticipate the user's needs in specific situations and ask proactively. Although voice control is becoming increasingly powerful, users are not clear about its limits. This is considerably different from on-screen interactions. If a new function is added on-screen, users will see a new icon. When Apple's voice assistant, Siri, is activated without a usage guide, many users will not know what it can do.

The introduction of autonomous driving, including driver-assist features, has altered the traditional paradigm in which a vehicle is solely controlled by the driver, causing many drivers to feel uneasy when the vehicle is not under their control. Therefore, even though autonomous driving does not require frequent interaction with the driver in terms of working mechanisms, it should still present the surrounding conditions in real time on the screen to assure the driver that it is "constantly monitoring" and "working reliably." Tesla Model Y can even identify whether the surrounding vehicles are sedans, SUVs, buses, or motorcycles, and display this information on

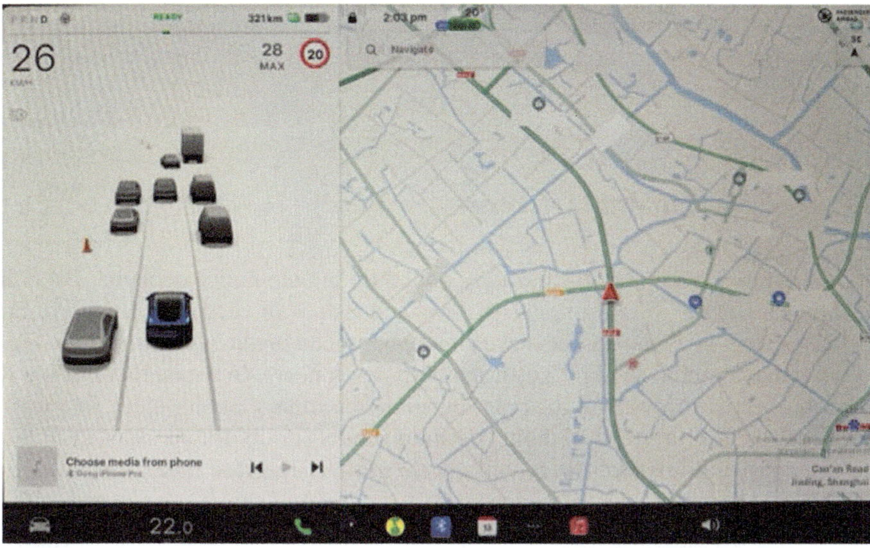

Fig. 10.12 Surrounding environment displayed on Tesla Model Y (2023)

the central information display, as shown in Fig. 10.12. Although categorizing the surrounding vehicles may not directly aid the strategy of the autonomous-driving algorithm, it can make users feel that the vehicle's recognition capability is very powerful, thus building more trust in the reassurance of autonomous driving.

The values of reassurance and surprise and delight might appear contradictory; however, they have different application domains. Reassurance primarily applies to functions with a clear purpose, especially when users are not accustomed to their interaction modalities. By contrast, surprise is primarily applied to features beyond the users' expectations, and the presence or absence of these features does not affect the primary functions of the HMI system. Therefore, it is absolutely possible for a vehicle to simultaneously provide both reassurance and surprise and delight.

10.3 Other Reasons for Differences in Automotive User Experience

The research methods for culture and values are not only applicable to automotive HMI systems but can also be employed in several other areas such as overall automotive experience design, product definition, and marketing strategies. This is especially true when automotive companies venture into international markets, where understanding the culture of the local market is vital for the success of their products [6].

However, cultural differences are not the only reason for differences in automotive user experience. We should not attempt to explain all differences in user experience with cultural research, even if we have mastered the comprehensive methodology. In addition to culture, differences in user demands can be attributed to two major factors: generational and habit differences.

10.3.1 Generational Differences

Digitalization is becoming increasingly important in automotive cockpits. Users in different markets show varying onset times and speeds of accepting digitalization. Thus, the users of some markets have a more advanced understanding of digitalization in the automotive cockpit, resulting in more pioneers. Conversely, the users in some markets are slower at grasping digitalization, thus yielding more followers. However, the general trend of digital development is similar for all users, such that the demands of digital followers will develop in the direction of digital pioneers, albeit with a time lag.

The influence of the Internet is a significant factor in creating this generational difference. China is among the countries where the Internet has spread the fastest worldwide. In 2022, the market share of 5G smartphones in China was approximately 84%, with an average daily cell phone usage of up to 3.3 h per person. Since 2016, the "cashless" lifestyle has become increasingly prevalent in Chinese cities. Most daily transactions are completed via smartphones, eliminating the need for banknotes, coins, or physical bank cards. This heavy reliance on smartphones and the Internet has also made users more familiar with the interaction paths, hierarchical logic, and menu naming of smartphones. Therefore, transferring interactions similar to those on smartphones to vehicles or other home appliances is very easy to grasp for many users.

Furthermore, the intelligent vehicle industry is thriving in the Chinese market. Since 2018, more than 50 new automotive brands have been established in China, and traditional car manufacturers have been constantly innovating. For most automotive brands, the intelligent cockpit is among their most important selling points. This has enabled Chinese users to experience more, better, and newer HMI systems, which has cultivated their usage habits and raised their expectations.

Additionally, the Chinese automotive market had a relatively late start. In 2019, more than 50% of consumers in the Chinese market were first-time car buyers, meaning that most Chinese users have no experience with traditional automotive HMI systems (i.e., car models produced between 2000 and 2015). They have never tried controlling a non-touchable central information display with a knob and searched for information on a small 5- or 7-in. central information display. Therefore, automotive HMI interfaces inherited from traditional designs are not appealing to Chinese users but may appear strange and outdated.

For these reasons, the design of automotive HMI systems in the Chinese market exhibits a considerable generational difference. The unique design characteristics resulting from these generational differences and not driven by cultural differences are expected to be gradually followed in other markets. For example, China's NIO ES8 began using a central information display interaction design with maps and widgets as the homepage as early as 2018, whereas Mercedes-Benz started using a similar homepage design, namely, the MBUX Zero Layer, in 2021.

10.3.2 Habit Differences

In addition to cultural and generational differences, differences in user habits are also reasons for varying needs in user experience design. Some habits stem from cultural differences; however, the corresponding experience demands are not directly caused by culture. Some typical habit differences include the following:

Differences in the family structure: compared with European users, a significant proportion of Chinese users live in three-generation households. Therefore, in addition to daily commuting, a vehicle often needs to pick up and drop off children and the elderly, which increases the importance of scenarios involving rear passengers, children, and elderly individuals.

Expression characteristics of Chinese characters: Chinese characters are block-shaped, and most words are composed of 2–3 characters. The text for options on the central information display usually does not exceed five Chinese characters; therefore, the corresponding icon cards do not need to be very wide. However, languages such as English and German are made up of alphabetic words, and the length of each word varies greatly, requiring icon cards to have sufficient width to accommodate words of different lengths. Therefore, if only the typesetting of Chinese pages is considered, the interface layout will have higher flexibility.

Frequently used smartphone applications: specific frequently used smartphone applications can also shape user habits. For example, Western users often use Google applications; therefore, they are accustomed to Google's interaction logic and page layout. However, the majority of Chinese users do not use Google applications; instead, they use local applications such as WeChat and AliPay, which will affect their habits accordingly.

References

1. Geert Hofstede, Gert Jan Hofstede. Cultures and Organizations: Software of the Mind (2nd ed.) [M]. Translated by Li Yuan and Sun Jianmin. Beijing: Renmin University of China Press, 2010.
2. Geert Hofstede, Gert Jan Hofstede, Michael Minkov. Culture and Organizations: Software of the Mind (3rd ed.) [M]. McGraw-Hill, 2010.
3. Christian Schmidt. Uncertainty in Economic Thought [M]. People's Publishing House, 2020.
4. Huib Wursten. The 7 Mental Images of National Culture [M]. Hofstede Insights, 2019.
5. Jean-Pierre Coene, Marc Jacobs. Cross-border negotiation localization strategy [M]. Fudan University Press, 2019.
6. Liviu Warter, Lulian Warter. Intercultural Issues in the Global Auto Industry. Nova Science Publishers, 2021.

Chapter 11
Aesthetics

11.1 Development

11.1.1 Aesthetics in Interface Design

The aesthetics appeal of automotive HMI systems is another important factor that can affect the overall user experience. Compared with other evaluation indexes, the evaluation of aesthetics is usually more intuitive and quick to judge. When consumers go to a showroom to purchase a car, their first impression when entering the cabin includes the attractiveness of the interface on the central information display. This initial impression is crucial for consumers when deciding whether or not to purchase the car. Therefore, automotive brands must ensure the aesthetics of the system interface to win the consumers' favor. In fact, aesthetics has always been a part of the product's core value, not only in automotive HMI but also in a wide variety of consumer products. Humans possess an innate sense of esthetic appreciation. As posited by Immanuel Kant in his *Critique of Judgment* published in 1790, the year after the French Revolution, aesthetics is a universal and shared aspect of the human's spiritual world. The aesthetic evaluation index stems precisely from this notion.

The aesthetics index is intended to evaluate the visual attractiveness of the user interface design. The user interface is an important factor influencing user experience. Tractinsky's experiment on automated teller machine interfaces in 1997 revealed a strong correlation between the users' perception of system quality and their evaluation of interface aesthetics [1]. The formal inclusion of interface aesthetics within the scope of user experience occurred in 2007 when Thüring proposed the components of user experience (CUE) model [2]. In this model, the CUEs are identified as the perceptions of instrumental qualities, emotional reactions, and non-instrumental qualities, within which visual aesthetics is also included, as shown in Fig. 11.1. In addition, numerous studies have reported that the esthetic appeal of the product contributes significantly to overall user satisfaction. Individuals tend to be more

J. Ma and Z. Gong, *Automotive Human-Machine Interaction (HMI) Evaluation Method*, Research on Automotive Intelligent Cockpit,
https://doi.org/10.1007/978-981-99-9951-4_11

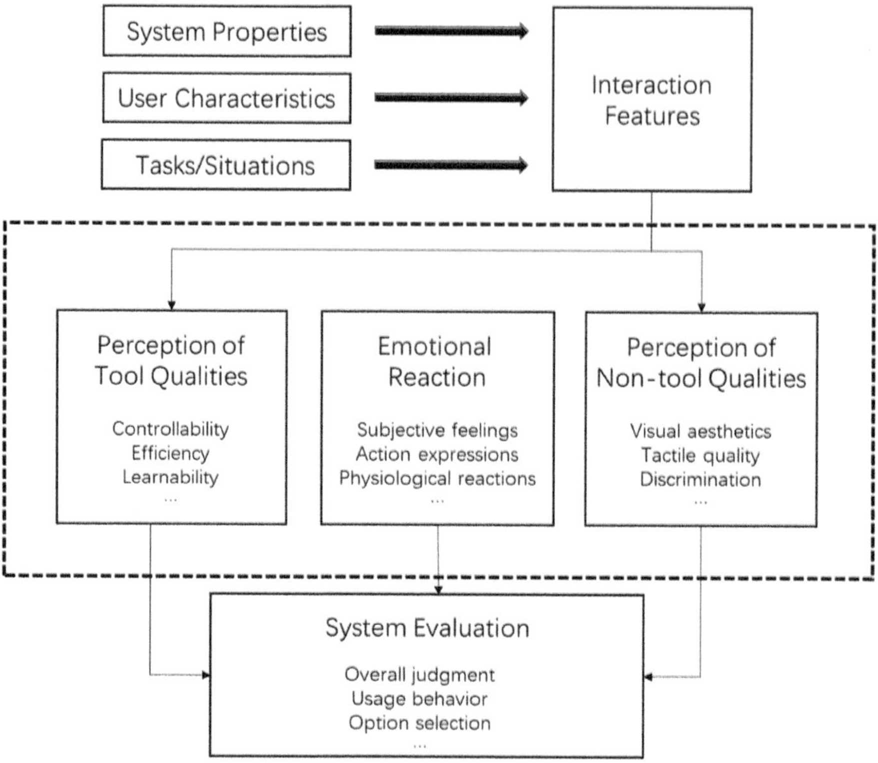

Fig. 11.1 CUE model for the elements of user experience (*Source* Thüring, 2007)

satisfied with products that are visually appealing but less user-friendly than with products that are easy to use but less aesthetically pleasing. For example, Cyr et al. conducted aesthetics-related experiments on mobile websites using the technology acceptance model; their results showed that the esthetic quality of interfaces in interactive products has a significant impact on the users' perception of system usability, ease of use, and enjoyment [3].

The history of interface design reflects the constant pursuit of higher aesthetic levels in interfaces. In the 1970s, researchers at the Xerox Palo Alto Research Center developed the first graphical user interface (GUI), ushering in a new era of computer graphical interfaces. Since the 1980s, the interface design of operating systems has undergone numerous changes, evolving from less esthetically appealing character-based interfaces to more visually attractive GUIs. Microsoft's Windows 95 operating system in the mid-1990s represented the shift from character-based to graphics-based interfaces. Currently, interface design is evolving towards greater humanization and personalization. Visual elements such as images, patterns, and colors enable the interface itself to display the corresponding esthetic and humanistic values, which also allows the expression of increasingly rich symbolic meaning.

Fig. 11.2 Left: Nokia 3650 cell phone (2002); Right: Nokia 7610 cell phone (2004) (*Source* Nokia)

User interfaces can be classified into physical and virtual interfaces. Physical interfaces include the screen position, button location, and design, which are generally presented in a three-dimensional space. By contrast, virtual interfaces refer to the content displayed on the screen and are primarily presented on a two-dimensional plane. Currently, the design of various consumer products places a greater emphasis on virtual interfaces. This is because users' aesthetic perception of interactive products has shifted from overall appearance to the on-screen interface. More than a decade ago, cell phones relied on their exterior design and innovation to attract consumers' attention. Even model iterations from the same brand would emphasize distinctive exterior designs. By contrast, the interface design underwent minimal changes, as shown in Fig. 11.2. This phenomenon has essentially disappeared in today's cell phone market to the extent where distinguishing between different brands based on exterior appearance alone is often difficult for consumers. However, the richness of current smartphone interface design is unprecedented. For example, as shown in Fig. 11.3, two smartphones can have similar appearances, differing only in minor details; however, their interfaces are stylistically different, which can be used to attract consumers' attention.

The aesthetic evaluation of automotive HMI systems discussed in this chapter will primarily focus on the design aesthetics of virtual interfaces. This is partly because virtual interfaces are becoming the central features of cockpit design, with relatively little distinction in physical interface designs among different car models. Besides, it is difficult to develop standardized evaluation methods in three-dimensional space.

11.1.2 Aesthetic Evaluation

Aesthetic evaluation, an entry in the *Dictionary of Psychology*, refers to the process during aesthetic activity in which a subject makes value judgments regarding the

Fig. 11.3 Left: VIVO X60 smartphone (2020); Right: HUAWEI P50 smartphone (2021) (*Source* VIVO and HUAWEI, respectively)

attributes of an aesthetic object based on their unique aesthetic values [4]. It is influenced by aesthetic cultivation, intellectual level, and personal preferences, among other factors. Some young users with flamboyant personalities who were born during the Internet era might consider an HMI interface with vibrant colors and a multitude of elements to be appealing, whereas elder users who are more mature and composed might prefer a more minimalist interface.

Although aesthetic evaluation is largely subjective, it is not boundless, especially in the context of automotive HMI systems. Instead, it follows certain principles. Symbolism can provide some insights for the aesthetic evaluation of automotive HMI systems. Originating in the late 19th century in the United Kingdom and several other Western countries, symbolism is a school of thought dedicated to transforming abstract concepts into tangible objects. It uses items from the objective world to represent the subjective world, whereas the objective items are de-emphasized. As the symbolist Gustave Kahn said, "The essential aim of our art is to objectify the subjective (the externalization of the idea), instead of subjectifying the objective (nature seen from the eyes of a temperament)." The theoretical foundation of symbolism is science, that is, to identify the underlying laws of phenomena. Consequently, the corresponding relationships between ideas and things are traceable and relatively fixed. In line with these ideas, researchers proposed certain connections between the properties of color and lines and the human subjective world. Following the emergence of this theory, it is believed that the laws of beauty exist in the external objective world. Thus, inner emotions can be translated into tangible objects, colors, or shapes. This is akin to encoding human thoughts and emotions into a series of characters such that they can be scientifically measured and expressed.

In aesthetic evaluation, a large part of what constitutes "appealing" leans toward the symbolic. Therefore, by employing the ideas of symbolism, these elements can be evaluated in a more scientific and standardized manner while also aligning more closely with the aesthetic habits of most users. When considering aesthetic attributes, such as color schemes and lines, in interface design, the focus is on whether the specific manifestation of this attribute symbolizes a certain subjective impression. For example, most users would agree that the color red in a vehicle's instrument cluster display represents speed and performance, without overly concerning themselves with the beauty or ugliness of the red element. This approach is more feasible and standardized compared with the direct evaluation of interface aesthetics. Thus, whether a certain color scheme or pattern aligns with users' purely subjective and symbolism-unrelated preferences for beauty will not be evaluated in this book. At the current stage of automotive HMI development, enhancing the users' perception of certain vehicle attributes through interface design and conveying intrinsic meanings and imaginations that transcend appearance is critical. For instance, in the shutdown animation presented on the instrument cluster display of the Mercedes-Benz S-Class sedan, a dense array of three-pointed stars twinkles around the logo, symbolizing the vehicle's luxury and grandeur, while the rendering of a metallic texture symbolizes machinery, which conveys a sense of technology and future to the users, as shown in Fig. 11.4.

In summary, the aesthetic evaluation of automotive HMI systems can be conducted by employing the following standardized process: first, several trends in virtual interface design are summarized. Second, typical symbolic design approaches corresponding to each style are identified to serve as specific checklist items. Third, whether the car model under evaluation employs these design approaches and the extent to which these approaches are applied are examined, that is, the quantitative scoring of each item. Finally, the level attained by the evaluated car model for each design style is assessed.

The scores for different styles are not cumulative. An HMI system can achieve a relatively high level in several aesthetic directions or it can reach the extreme in

Fig. 11.4 Shutdown animation on the instrument cluster display of the Mercedes-Benz S-Class (2020)

only one direction. We cannot claim that the former design is superior to the latter but that each has its distinctive orientation. In other words, we can only determine which system performed better within a specific aesthetic direction.

The advantage of this method lies in its standardization and quantification of aesthetic evaluations, however, it has two limitations. First, this approach cannot fully encompass all elements of aesthetic evaluation but focuses primarily on symbolic elements or elements that can be readily abstracted and assessed. Second, although this method can guide designers to learn from the strengths of current benchmark products to produce high-quality interface designs, it does not inherently lead to innovation; hence, it cannot play a role in the production of disruptive, exceptional products. Nevertheless, fostering innovation is not the main goal of any evaluation system.

11.2 Major Aesthetic Styles

In automotive HMI, the style of the virtual user interface design is created by integrating specific design elements, such as color, graphics, and layout. This combination forms representative features on the interface and evokes specific overall impressions in users. Currently, the major styles of interface design in automotive HMI systems include the following: sense of luxury, simplicity, sense of technology, and warmth.

11.2.1 Sense of Luxury

Sense of luxury refers to the sense of quality, opulence, and extravagance imparted by the interface design.

The pursuit of luxury has been a constant theme throughout the history of art, with the Baroque and Rococo styles in the 17th to 18th centuries, respectively, serving as quintessential examples. The term "Baroque" originates from the Portuguese word *barroco*, which means "an irregularly shaped pearl". It is an art style characterized by grandeur and ornateness that began in the 17th century. "Rococo" is a combination of the French term *rocaille* ("shell art") and the Italian term "*barocco*" (Baroque), with some considering Rococo as the late phase of the Baroque style. Both styles are characterized by lavish decoration, and their representative works meticulously depict every intricate and exquisite detail to satisfy the nobility's cravings for luxury. For instance, the ceiling fresco *The Triumph of Divine Providence* is one of the earliest and most popular examples of Baroque painting, as shown in Fig. 11.5.

Artists and their works have associated luxury with intricacy and refinement, the combination of which is indicative of high quality. Whether in architecture or painting, intricacy and refinement require a greater investment of time and money, and the pursuit of high quality through increased cost constitutes luxury. As time passed

Fig. 11.5 *The Triumph of Divine Providence* (detail) by Pietro da Cortona (1633)

and people's tastes evolved, the direct association between the sense of luxury in contemporary interface design and the opulence in classical art weakened significantly. However, the aspiration for luxury remains constant; hence, the core idea has been inherited and reflected in interface design through modern concepts.

High-quality images, highly saturated contrasting colors, and tangible textures are elements that create a sense of luxury. These are also associated with expensive hardware. All designers use these high-quality elements in their pursuit of ideal effects when designing interfaces. However, several obstacles exist between the design draft and the actual presentation, including screen quality and computing power. For a single interface, high-quality images with tangible textures require high-resolution screens, while elements with high saturation require the screen to have strong color reproduction capabilities. For example, in the iDrive 8 operating system launched by BMW in 2021, the metallic texture created using rich color gradients and changes in light and shadow is presented on a screen with a resolution of approximately 200 PPI to forge a sense of luxury for users, as shown in Fig. 11.6. For the overall system, switching between interfaces occurs very frequently; hence, rendering the content of each high-quality interface within a very short time poses a certain level of challenge to the computing power of the in-vehicle chips.

Curves and textured embellishments are design elements that can also convey a sense of luxury. Works in the Baroque and Rococo styles make good use of delicate and slender curves and spirals, often drawing on textures found in nature, such as the use of vines as decorations to embody delicacy, intricacy, and luxury. In contemporary interface design, curves and textures are also employed to make intricate and refined designs. Compared with straight lines, curves are more flexible and livelier but are also more difficult to control. Minor changes in the curvature can impact the overall effect. Designers need to exercise more delicate control when using curved lines to design interfaces, and will naturally present more details in the interface effect. Delicate textures that enrich interface details and interface layering also play a role. The aforementioned BMW iDrive 8 interface uses triangular textures to add a sense

Fig. 11.6 Instrument display interface and partial magnification of the BMW's iDrive 8 system (2021) (*Source* BMW Group)

of layering to the material (Fig. 11.6). Although it does not cover a large area and is less easily perceived by the user at first glance compared with large areas of light and shadow, such exquisite details demonstrate the designer's meticulous attention to detail, further enriching the user's sense of luxury and quality.

11.2.2 Simplicity

In interface design, simplicity refers to the reduction of unnecessary decorations while retaining and refining core elements.

As human civilization progresses and social forms change, the pursuit of luxury and decoration is no longer the mainstream aesthetics for people. After two industrial revolutions, modernist design emerged as a new trend. This movement was opposed to extravagance and emphasized geometrization, simplicity, and industrialization, almost in diametric opposition to traditional luxury. The essence of modernist design is functional, rational, anti-traditional, and anti-decorative. It not only established new forms and aesthetics for the designs of mass-produced products but also profoundly influenced subsequent architectural, environmental, graphic, and interior designs. One of the most important proponents of modernism was Ludwig Mies Van der Rohe, the third director of the Bauhaus School of Design. His motto, "Less is more," is still revered and has deeply ingrained the minimalist design style into the collective consciousness.

Modernism emphasizes function as the center and purpose of design, whereas form is secondary. Thus, it advocates for simple, direct, and efficient designs that are user-friendly. This concept has had a significant impact on the field of graphic design. For instance, many poster designs since the early 20th century have reduced intricate decorations, retaining only core essential information, as shown in Fig. 11.7. Currently, with the development and popularization of smart devices, interface design has gradually separated from graphic design to form a relatively independent field. However, the modernist notions of function-centered and simplistic designs have continued to influence interface design.

The simplistic style has become one of the mainstream styles in interface design for two main reasons: first, the majority of contemporary users are accustomed to the interface interactions of smart products and are familiar with general operating methods and graphic connotations. Therefore, icons on the interface do not need to mimic the concrete, original appearance of real things. Instead, simple and abstract flat icons can directly convey their essence, enabling users to understand their meaning. For example, the icons in Apple's smartphone operating system have become increasingly simplistic and abstract over the past decade, as shown in Fig. 11.8. Second, simplistic interfaces can reduce the users' cognitive load and enhance readability and legibility. With no redundant and irrelevant information occupying the users' visual perception, they will be able to read relevant information

Fig. 11.7 Poster for a Bauhaus exhibition by Joost Schmidt (1923)

Fig. 11.8 Icons in the iOS 5 (2011) and iOS 15 (2021) operating systems

at first glance. Furthermore, interference during processing in the brain is also minimized, thus reducing the information processing workload. In the current fast-paced lifestyle, this also aligns with people's pursuit of efficiency.

Simplistic interfaces often replace color blocks with large areas of white space, and usually feature black, white, and gray as the main colors. The substitution of decorative elements with white space serves to highlight the main content and reduce

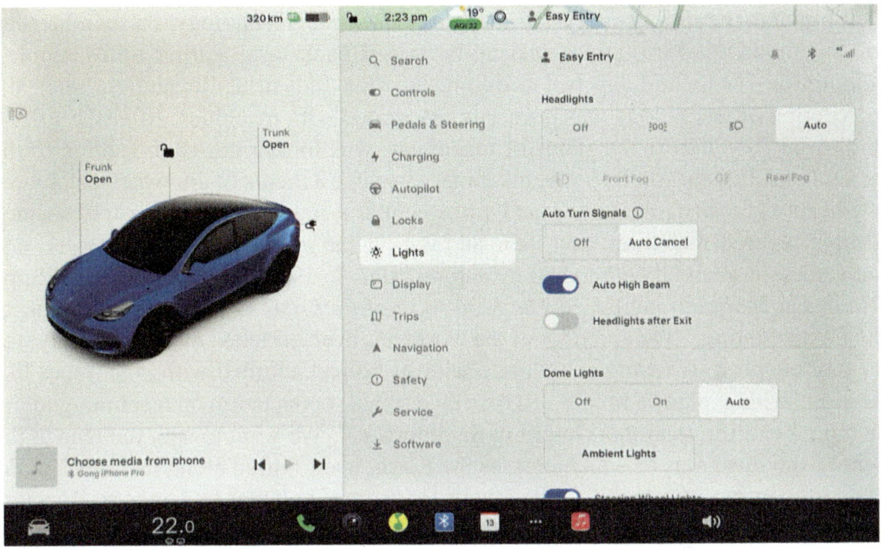

Fig. 11.9 Settings interface of the Tesla Model Y (2023)

irrelevant information in the interface. Furthermore, colors have rich meanings, which individuals tend to unconsciously interpret. Employing a monochromatic main color scheme for the interface can significantly reduce the delivery of redundant information to users, allowing information to be predominantly conveyed through shape and layout. However, in certain situations, the role of colors is irreplaceable. In such cases, using color diffusion and gradients can substitute the role of colors to a certain extent, conveying comprehensive information to users without undermining the overall simplistic style. In addition, simplistic interfaces often employ abstract and minimalist line icons. In the central information display interface of Tesla's Model Y, all icons are presented with the most simplistic design to convey meanings and functions, as shown in Fig. 11.9. Their categories and functions are primarily distinguished by position rather than complex borders, and a gray scale is utilized to partition large areas. Moreover, in terms of the overall layout and color usage, the interface design of the Model Y uses extensive white space and a monochromatic main color scheme to enhance the users' perception of the simplistic interface style.

11.2.3 Sense of Technology

In recent years, the sense of technology in interface design is primarily elicited by digital elements that evoke a sensation of transcending reality. However, unlike the previous aesthetic styles, the definition of a sense of technology is not fixed, as technology itself is constantly evolving and changing.

Similar to the simplistic style, the origin of a sense of technology is also influenced by the Industrial Revolution and can be traced back to an earlier time. People's admiration for technology began during the first Industrial Revolution when the perception of technology was primarily associated with machines. After the second Industrial Revolution, the status of machinery was further elevated, leading to the advent of futurism. At this point, machinery became a theme of art, which celebrated the beauty of speed, machinery, and movement associated with an industrial society. People revered machines, and their admiration for machinery, speed, power, and movement was vividly embodied by automobiles. In 1909, the Italian writer Filippo Tommaso Marinetti published *The Manifesto of Futurism* in the French newspaper *Le Figaro* stating, "The splendor of the world has been enriched by a new beauty: the beauty of speed. A racing automobile with its bonnet adorned with great tubes like serpents with explosive breath… a roaring car that seems to run on machine-gun fire is more beautiful than the Victory in Samothrace… We want to sing the man at the wheel, the ideal axis of which crosses the Earth, itself hurled along its orbit." In the field of painting, although the 1912 work *Horizontal Volumes* by Umberto Boccioni depicts a human figure, it conveyed the speed and power of machinery via distorted facial features and clothing in disarray, as shown in Fig. 11.10. The complexity and robustness of automobiles instilled a sense of awe towards machinery. In the early days of automobiles, if a vehicle hit a pedestrian and caused a traffic accident, the blame was generally placed on the pedestrian for failing to avoid the vehicle.

Since the advent of the third Industrial Revolution, humanity has been undergoing a phase of vigorous technological development, with the importance of technology constantly growing. In today's world, science and technology are the most active factors driving productivity across virtually all sectors of society, and the automotive industry is a prime example. Modern automotive manufacturers have placed increasing emphasis on advancing and communicating their product technology, while a "sense of technology" is a perceptual means by which manufacturers can convey their products' technological sophistication to users. For most nonprofessional users, the sense of technology in a vehicle's exterior appearance or interactive interface design is more intuitive compared with various technical parameters, possessing an irreplaceable appeal. Furthermore, in-vehicle screen hardware is a form of technological product, providing an excellent foundation for the virtual HMI interface to convey a sense of technology.

The emergence and advancement of electronic information and the Internet have shifted people's perception of technology from tangible machinery to virtual digital elements that transcended the machines from the speed of machinery to computational speed, and from engine power to information-processing power. In the automotive HMI interface, designs that incorporate the abstraction and digitization of contemporary real-world technological elements would inherently possess a sense of technology. For instance, the simulation of microscopic particle arrangements, line designs reminiscent of circuit boards, and various styles of light and shadow effects essentially reflect technological elements comprehended by individuals in their everyday lives. Additionally, by forging a holistic spatial three-dimensionality within the interface and integrating clear-cut, smooth animations can enhance the

Fig. 11.10 *Horizontal Volumes* by Umberto Boccioni (1915)

users' perception of technology. When conveying a sense of technology, interface design usually employs color schemes dominated by dark tones such as green and blue. Early technology-oriented interfaces commonly used green, most likely attributable to the long-standing use of single-color green monitors as the primary output device for computers. Examples include the Apple Monitor III in 1980 and the IBM 5151 Monitor in 1981, which were only capable of displaying green. It was not until the late 1980s, when the Microsoft DOS operating system reached maturity, did black and white and color displays gradually became more common. Green has also appeared in some works of art, such as *The Matrix* film trilogy (1999–2003), which extensively utilized a graphic style with black backgrounds, green text, and silhouettes, profoundly influencing subsequent aesthetics in the sense of technology (see Fig. 11.11). In recent years, blue has begun to replace green as the mainstream representative color for conveying a sense of technology, such as in the interface of the Mercedes-Benz EQE SUV, as shown in Fig. 11.12. The shorter wavelength and higher color temperature of blue are apt for expressing the calmness and rationality unique to machines. Additionally, blue is a relatively rare color in the natural biological world and hence is more suitable for expressing elements projected from the virtual world into the real world.

Fig. 11.11 Album cover for *The Matrix 2: Reloaded* original motion picture soundtrack (2003)

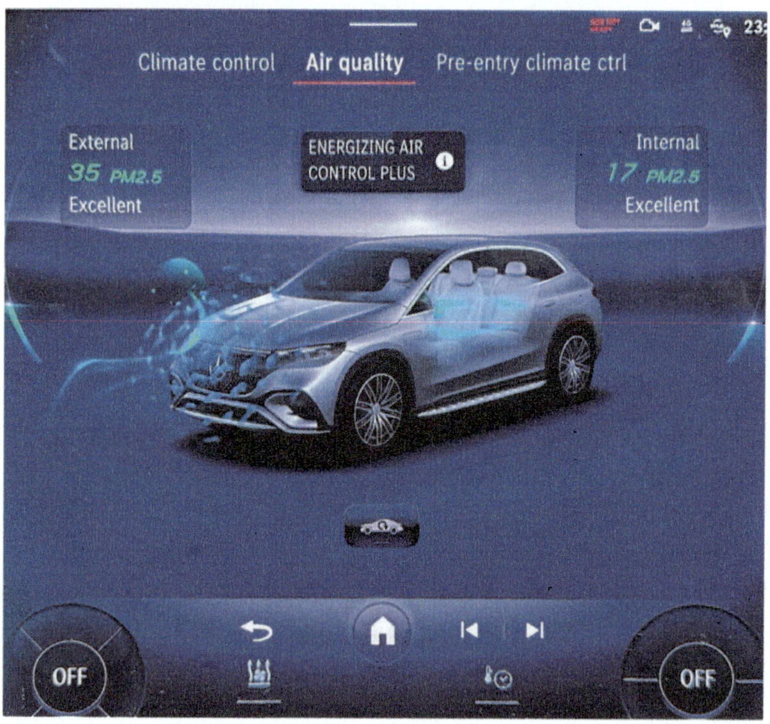

Fig. 11.12 Air quality interface of Mercedes-Benz EQE SUV (2023)

11.2.4 Warmth

Warmth refers to a human-centered concept utilized by technology products to convey humanitarian care, counteracting the inherent coldness of technology. Recently, the HMI design of several Chinese automotive brands has leaned towards embodying this design aesthetics, whereas mainstream products from German brands rarely demonstrate this particular aesthetic direction.

Warmth and coolness are among the most fundamental human perceptions. However, with the development of modern society, the term "warmth" has transcended the scope of physiological perception. In the field of design, temperature embodies human-centered care. A "warm" design places human feelings at its core, emphasizing the idea that technology is in service to humanity. This design philosophy contrasts with the notion of sheer reverence for technology. Instead of directly emphasizing technology itself, it allows technology to recede and hide beneath visual representations. Warmth does not imply anti-technology. Instead, it brings technology closer to people's hearts in a gentler manner by demonstrating the usage scenarios of technology in people's lives.

In interface design, "warm" interfaces often utilize light or warm colors as their main tone, and grayscale is usually incorporated to a certain extent when applying colors to maintain a clean and clear interface. For example, in its dark mode, the HMI system of NIO ES8 employs a gradient orange color to emphasize key information, reminiscent of the warm hues of dusk, thereby invoking feelings of relaxation and warmth, as shown in Fig. 11.13. Additionally, buttons, cards, and other graphical elements in these "warm" interfaces tend to deliberately avoid sharp angles, opting instead for rounded designs. This is partly because the physiological structure of the human eye finds it easier to identify curved shapes with rounded corners and is also guided by rounded corners to focus on the center. Furthermore, the graphical attributes of rounded corners are softer and more comfortable, providing a sense of safety and intimacy, whereas sharp angles can feel cold. In addition to the design considerations described above, some interface designs employ illustrative elements or micro-interaction effects to create warmth. The popularity of "warm" vehicle interface designs underscores the fact that the fundamental purpose of technology is to enhance the quality of human life.

In addition to the four major aesthetic styles discussed here, other styles can be utilized by some brands or car models to emphasize their unique product attributes. For instance, the interface of the French brand DS boasts a sharp and haughty style reminiscent of the Louvre Pyramid, whereas the interface of the British brand MINI features a round and lively style. However, these styles are not representative and therefore are not included in this discussion.

Designers can use these interface styles and their respective common design approaches as a reference to optimize interface design and enhance the symbolic meanings conveyed by the interface. In addition, they should utilize their abilities to fully integrate various typical design approaches. However, using these styles

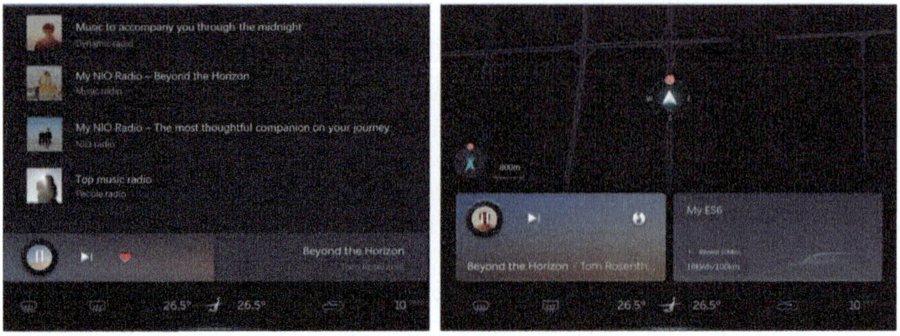

Fig. 11.13 Partial interface of the NIO ES8's central information display (2020) (*Source* NIO Inc.), translated from Chinese language

as a reference should not limit designers; exceptional disruptive ideas can potentially create successful products, even without adopting the aforementioned design approaches.

References

1. Tractinsky, Noam. Aesthetics and apparent usability: Empirically assessing cultural and methodological issues [C]. Proceedings of the 1997 Conference on Human Factors in Computing Systems, CHI, 1997.
2. Thüring M, Sascha M. Usability, aesthetics and emotions in human-technology interaction [J]. International Journal of Psychology, 2007, 42(4): 253–264.
3. Cyr D, Head M, Ivanov A. Design aesthetics leading to m-loyalty in mobile commerce [J]. Information & Management, 2006, 43(8): 950–963.
4. Lin Chongde. Dictionary of Psychology [M]. Shanghai: Shanghai Education Press, 2003.

Chapter 12
Application of the Evaluation System in the R&D Process

12.1 Application of the Three-Dimensional Orthogonal Evaluation System

In this book, we introduced a three-dimensional (3D) orthogonal evaluation system for the automotive HMI and provided an in-depth analysis of its evaluation indexes. In this section, we offer relevant suggestions on how this evaluation system should be applied in the actual development process of an automotive cockpit.

12.1.1 Suggestions for Evaluation Item Distribution

In the 3D orthogonal evaluation system for an automotive HMI, each evaluation item is situated within a 3D space constructed by the evaluation indexes, interaction tasks, and interaction modalities, as shown in Fig. 12.1. Theoretically, the number of evaluation items is the product of the quantity of the three dimensions, which could reach tens of thousands. However, during the actual testing and evaluation process, some of these theoretical evaluation items are not important. For example, complex and infrequently used vehicle settings are usually operated only when the vehicle is parked. Therefore, it is not necessary to examine their safety during the driving process. Furthermore, some items are inherently meaningless. For instance, the evaluation of anti-noise performance in utility is only effective for tasks using the voice control modality, whereas operations involving touch screens and buttons are clearly not affected by noise. Hence, these two types of evaluation items do not need to be included in the actual evaluation system. The specific selection process can be adjusted according to the actual project requirements.

For comprehensive automotive HMI testing and evaluation projects without a specific focus, we propose a typical evaluation item distribution, as shown in Table 12.1. The classification of interaction tasks is detailed in Chap. 3, Sect. 3.2.

J. Ma and Z. Gong, *Automotive Human-Machine Interaction (HMI) Evaluation Method*, Research on Automotive Intelligent Cockpit,
https://doi.org/10.1007/978-981-99-9951-4_12

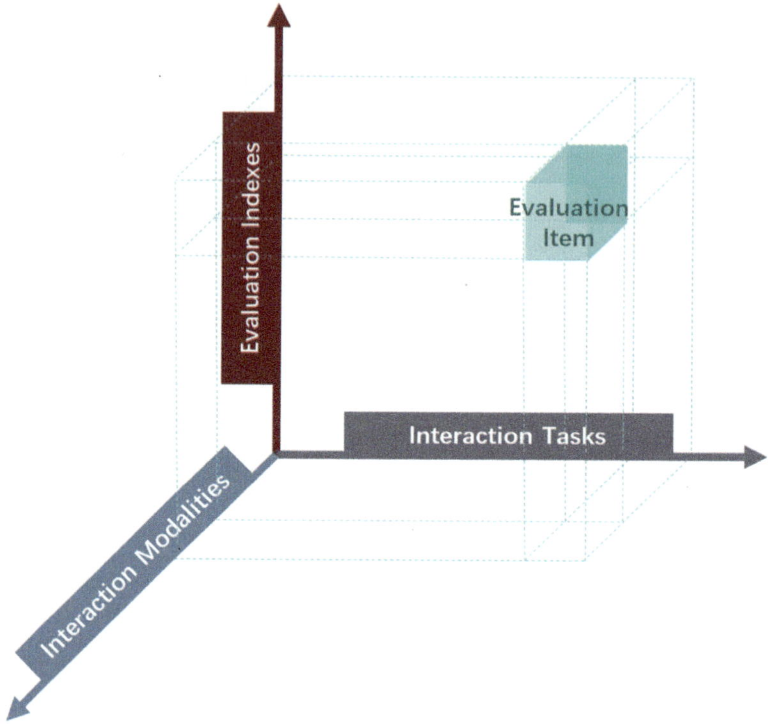

Fig. 12.1 Structure of the three-dimensional orthogonal evaluation system

The test result of each evaluation item can be converted into a score according to the corresponding measurement standards. Subsequently, the scores of all items are summed according to their weights along the three dimensions to obtain the total evaluation score.

12.1.2 Score Processing Methods for Multiple Interaction Modalities

When calculating the total evaluation score, weighted summation is a unified method that provides intuitive results for the evaluation index and interaction task dimensions. However, several methods can be used for processing the scores of the same task completed using different interaction modalities.

The first method of processing the interaction modality score is to average the different interaction modalities under each interaction task. This method appears very intuitive and comprehensive; however, it may restrain the number of interaction modalities for each task. For example, a certain interaction task can score up

Table 12.1 Proposed distribution of automotive HMI evaluation items

Evaluation indexes		Interaction tasks					
First-level	Second-level	Basic interaction tasks	Extended interaction tasks	Ecological and scenario tasks	Basic system experience	Information reading	Navigation performance
Utility	Availability	●	●		□	■	■
	Task success Rate	●	▲				□
	Reachability	▲	▲				
	Stability	△			□		
	Modality enhancement	△	△		□		
Safety	Driving performance maintenance	●					
	Gaze diversion	●					
	Function restriction			□			
Efficiency	Task time	○			□		□
	Operation complexity	●	○				
Cognition	Logical structure	△	△		□		
	Element visibility	▲	▲		□	□	□
	Element understandability	△	△		□		□
	Element memorability	△	△		□		
	System feedback	△			□		
Intelligence	Comprehension				△		

(continued)

Table 12.1 (continued)

Evaluation indexes		Interaction tasks					
First-level	Second-level	Basic interaction tasks	Extended interaction tasks	Ecological and scenario tasks	Basic system experience	Information reading	Navigation performance
	Functional intelligence			□			
	Contextual intelligence			■			
Values	Companionship			□	□		
	honor			□	□		
	Surprise and delight			□	□		
	Reassurance			□	□		
Aesthetics	sense of technology				□		
	Sense of luxury				□		
	Simplicity				□		
	Warmth				□		

● Involves all tasks and all third-level indexes; each interaction modality is evaluated separately
○ Involves some tasks or some third-level indexes; each interaction modality is evaluated separately
▲ Involves all tasks and all third-level indexes; only applicable to certain interaction modalities
△ Involves some tasks or some third-level indexes; applicable to certain interaction modalities
■ Involves all tasks and all third-level indexes; unrelated to any specific interaction modality
□ Involves some tasks or some third-level indexes; unrelated to any specific interaction modality

to 9 with touchscreen and up to 7 with steering wheel button. If this task can be completed using both interaction modalities, the average score is 8. However, if the steering wheel button function is removed and only touchscreen interaction is available, the score would increase to 9. Thus, we can observe that the wider the range of interaction modalities supported, the less likely it is to obtain high scores under this scoring method. The second method involves considering only the best score for each interaction task and ignoring the scores of other interaction modalities. This method avoids the problem in the first method and does not lead to poorer scores for more diverse interaction modalities. However, it cannot be used to identify problems in other designs. For example, if an interaction task can achieve full marks with voice interaction, its perfect score will not be affected regardless of how poor its touchscreen interaction design is for this task. The third method is a combination of the first two methods, which involves assigning a higher weight to the best interaction modality and decreasing the weights of the second-best and subsequent interaction modalities. This method is more balanced; however, its operation is relatively complex; the process is not intuitive; and the results are not easy to interpret. Finally, the fourth method consists of assigning corresponding weights to different interaction modalities under each interaction task according to the usage frequency and based on market research inputs. When using this method, there may be differences in interaction habits between various target groups, and each user's interaction habits may evolve over time.

The choice of method for processing the interaction modality scores should be determined based on the specific evaluation objectives. If the goal is to identify problems in the design, the first or third method should be used. If the purpose is to maximize the potential of the product, the second or third method should be considered. If it is a one-time competitive benchmarking project, the fourth method can also be considered.

12.1.3 Limitations of the Evaluation System

The evaluation system introduced in this book aims to evaluate automotive HMI systems in a quantitative and standardized manner. Although values and aesthetics are two inherently subjective indexes among the seven primary evaluation indexes, it was still possible to standardize the evaluation of values and aesthetics to a certain extent by analyzing users' culture and symbolic techniques in interface design. By relying on this standardized HMI evaluation method, we can evaluate all mass-produced vehicles against a uniform yardstick to obtain quantitative and reproducible results, achieving a comprehensive, fair, and objective comparison among different products.

However, it is undeniable that standardized evaluation methods have certain limitations. First, this evaluation method cannot facilitate targeted in-depth research for certain innovative interactions. For instance, for directional speakers that can achieve independent sound zones, it is necessary to examine their level of sound zone independence and the level of background noise. These two indexes are meaningless for other interaction technologies and, therefore, would not appear in the standardized process of the evaluation method. Therefore, it would be necessary to specifically design targeted evaluation indexes for the evaluation of innovative interactions. These new indexes will only appear in the standardized process when a new interaction becomes more widespread in mass-produced car models.

Second, this evaluation method is not sufficiently detailed for examining users' purely subjective feelings. For instance, the level of pleasure users feel while using the system is not included in the evaluation system. This is because pleasure is affected by a wide variety of factors that are difficult to exhaustively enumerate. Moreover, these influencing factors are often not cumulative and are difficult to calculate, following the so-called "one flaw spoils all" principle. To study the users' purely subjective feelings in detail, subjective questionnaires should be used. However, owing to the poor reproducibility of subjective questionnaires, research results tend to only have strong comparability when conducted in the same period with the same interviewees.

Third, this evaluation method cannot differentiate variations in user expectations. For example, in the Chinese market, cars in the price range of approximately 20,000–60,000 Euros have the highest level of intelligence, and experienced users within this range also have higher expectations for automotive intelligence. However, users of entry-level products below 13,000 Euros and high-end products above 100,000 Euros may not be familiar with the most intelligent products on the market; hence, they do not have high expectations. Therefore, HMI systems with the same score may be considered excellent products for some users but not for others. However, from another perspective, the absence of subjective user expectations means that the results of this standardized evaluation method are more objective and fairer. In studies that need to incorporate users' subjective expectations, customer study could be used as a supplement to this evaluation method.

12.2 Testing and Evaluation Throughout the R&D Process

Product R&D in automotive HMI systems faces three main challenges. First, there is a wide variety of interaction architectures, and choices must be made. All cabin designers need to address the following questions: Should the central information display be horizontal or vertical? Should the central information display be placed above or below the air outlet? Is a front passenger display needed? Is a HUD needed? How many physical buttons should be retained? Should the home page of the central information display be a map or shortcut widgets? There is no absolute right or wrong answer to these questions, which implies that they all require analysis, verification, and proper selection. By contrast, the development goals of vehicle engines and chassis are clearer and more intuitive and, hence, are faced with fewer dilemmas.

Second, the long software development period makes it difficult to keep up with changes in user demands. Compared with software in smartphones and the Internet industry, the design and development of automotive HMI systems is slower, typically requiring 1–3 years. However, user demands evolve very quickly, and there are even cases where the product became obsolete upon market launch after development. For instance, in the Chinese market before 2019, watching online videos on the central information display was not a mainstream feature and was even ridiculed by some users. However, by 2021, the ability to play online videos had become an important indicator of automotive intelligence. In 2023, BMW's new-generation 5-series sedan (G60) launched in the European market with video-playing capabilities on the central information display.

Third, hardware is typically difficult to replace, which can limit software updates. The lifecycle of a vehicle can be as long as a decade or more, and vehicles are not replaced as frequently as smartphones. Although many intelligent vehicles can continually improve the software of their HMI systems via OTA updates, most central information displays and chips cannot be upgraded after the vehicle is purchased, thus limiting the performance of new software updates. Running the latest version on disparate hardware with variable specifications remains a challenge in software definition and development.

Traditional automotive testing and validation primarily targets fully developed, soon-to-be-released prototype products. The test results can be used to guide product fine-tuning and optimization, whereas major upgrades or redesigns of hardware and software architectures typically have to wait until the next generation of the model. Obviously, such a testing and validation process cannot fully address the challenges in automotive HMI product development. Testing and evaluation should be integrated into the entire process of automotive intelligent cockpit and HMI development. This includes rapid concept prototype validation after concept definition, high-fidelity prototype validation after UX/UI design, prototype module/prototype vehicle validation after system development, and mass-produced vehicle benchmarking after production. The testing and validation results for the output of each stage can be promptly fed back to the corresponding development team to avoid issues carrying on to the next stage and, hence, truly achieve rapid iteration, as shown in Fig. 12.2.

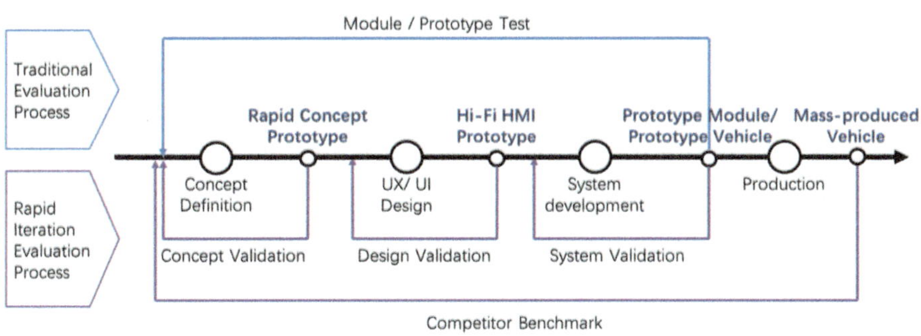

Fig. 12.2 Two different testing and evaluation processes in automotive product development

12.2.1 Concept Definition Phase

Most automotive cabin design teams aspire to create a system with unique highlights. Designers need to discern which ideas among a variety of innovative concepts will be favored by users. This cannot be achieved solely through the designers' imagination and reasoning; it also involves translating these ideas into tangible rapid concept prototypes, allowing both designers and potential users to enter an immersive experience.

The goals of testing and evaluating rapid concept prototypes are diverse and primarily depend on what new design concepts need to be validated. These may include the screen size and position, such as whether to use horizontal or vertical screens and whether to place them higher or lower; the number of screens, such as whether a front passenger display or a lower control display is needed; innovative interaction technologies, such as gesture control, touch-sensitive buttons, and directional speakers; definition of interaction modalities for crucial tasks, such as whether a touch screen can completely replace physical buttons for adjusting the climate control temperature; and layouts of key pages on the screen, such as whether to place maps or shortcut widgets on the homepage of the central information display. In addition, the users' subjective experiences when using the HMI system also need to be considered at this stage, such as whether an anthropomorphic voice assistant avatar would be more likely to gain users' trust than abstract waves. More innovative and distinctive designs require the investment of more resources for validation.

Rapid concept prototypes generally do not possess complete page elements or full functionalities; therefore, some evaluation indexes cannot be applied, and comprehensive testing and evaluation cannot be performed. Common evaluation indexes involved in concept prototype testing are listed in Table 12.2. One of the key points of validation at this stage is innovative interaction methods, including new interaction modalities and tasks, which can be categorized under availability and modality enhancement in utility. The overall page layout is related to some indexes in cognition. Other innovative concept designs may also involve evaluation indexes such as intelligence, values, and aesthetics. The evaluation of safety and efficiency usually cannot be conducted at this stage.

The goal of validation at the concept definition phase is to determine whether innovative concepts are accepted and appreciated by users, not the superiority or inferiority of specific design details. Accordingly, the test participants should primarily be real potential users. As their feedback is strongly subjective, the design of the evaluation process needs to be optimized to ensure that the evaluation results are as close as possible to the users' real attitudes. The rapid concept prototype should provide users a real experience in key functionalities, and the process design should also be close to real scenarios. For example, asking users directly "Do you think a dynamic, variable fragrance is useful?" may leave them at a loss for an answer, as they might not know where to start. A better process would be to first play the sounds and visuals of rain, switch the cabin's scent to a fresh, green fragrance, and then ask the user, "Do you need such a dynamic fragrance?".

Table 12.2 Evaluation index distribution according to importance in each R&D phase

Evaluation indexes		Design development deliverables				
First-level	Second-level	Rapid concept prototype	High-fidelity HMI prototype	Prototype module	Prototype vehicle	Mass-produced vehicle
Utility	Availability	● Innovative interaction methods	○	●	●	●
	Task success Rate			●	●	●
	Reachability			○	●	●
	Stability			○	●	●
	Modality enhancement	●	●	●	●	●
Safety	Driving performance maintenance				●	●
	Gaze diversion				●	●
	Function restriction		●	●	●	●
Efficiency	Task time				●	●
	Operation complexity		●	●	●	●
Cognition	Logical structure	●	●	●	●	●
	Element visibility		●	●	●	●
	Element understandability		●	●	●	●
	Element memorability	○	●	●	●	●
	System feedback	○	○	○	●	●
Intelligence	Comprehension	○	○	●	●	●
	Functional intelligence	○	○	●	●	●
	Contextual intelligence	●	●	○	●	●
Values	Companionship	○	●	●	●	●
	honor	○	●	●	●	●
	Surprise and delight	○	●	●	●	●
	Reassurance	○	●	●	●	●

(continued)

Table 12.2 (continued)

Evaluation indexes		Design development deliverables				
First-level	Second-level	Rapid concept prototype	High-fidelity HMI prototype	Prototype module	Prototype vehicle	Mass-produced vehicle
Aesthetics	Sense of technology	○	●	●	●	●
	Sense of luxury	○	●	●	●	●
	Simplicity	○	●	●	●	●
	Warmth	○	●	●	●	●

● Indexes primarily targeted in testing and evaluation
○ Indexes secondarily targeted in testing and evaluation

The seating buck for rapid concept prototypes typically consists of a simple cabin frame and a driving simulator, as shown in Fig. 12.3. The cabin frame contains seats and a steering wheel that closely approximate the size and position in a real car as well as screens for displaying the concept prototype, which are usually tablets. Participants operate the interfaces on the screens while driving in the simulator to achieve an immersive experience similar to a real driving environment. Testing of rapid concept prototypes can also be performed in a virtual reality (VR) environment. VR can replicate the full interior environment of a vehicle, resulting in a more realistic experience. However, even with 8K VR glasses, the angular resolution is still limited, making it difficult to clearly display smaller elements on the central information display and instrument cluster display. Thus, VR glasses are not suitable for tests that require users to read a large amount of small-sized text. However, with the advancement of technology, new virtual reality or augmented reality glasses like Vision Pro glasses released by Apple in 2023 will certainly bring more imaginative possibilities to the virtual concept test.

12.2.2 UX/UI Design Phase

Once the various concept definitions have been clarified, the user experience and user interface design (UX and UI design, respectively) phase begins. Prior to full-fledged software development following the completion of these designs, high-fidelity HMI prototypes can be developed swiftly and conveniently for design validation. For automotive HMI systems, the prototypes typically include all pages and elements in the screen interface and can operate according to actual interaction logic. Compared with the HMI system in real vehicles, high-fidelity prototypes have simplified animations and cannot communicate with vehicle electronic and electrical equipment. They usually run on computers or tablets rather than on real in-vehicle system hardware.

The testing and evaluation objective of high-fidelity HMI prototypes is to conduct a comprehensive and complete validation of the UX and UI design, including interface

Fig. 12.3 Rapid concept prototype seating buck

layout, interaction logic, visual style, element details, element sizes, and multi-screen linkage effects. Innovative points in the interaction design should be highlighted for validation at this stage. For instance, how well does the idea of sliding with multiple fingers to pull out a certain interface work in reality? Moreover, special attention should be paid in this phase to the relationship between screen positioning and the interface element layout. For example, when we consider the bottom area of a central information display, it would be easily reachable when the display has a higher installation position, whereas it might not be as easily accessible when the display has a lower installation position.

Compared with the indexes required for rapid concept prototypes, the evaluation indexes required in high-fidelity HMI prototype testing focus more on cognition while also including some efficiency and safety indexes, as shown in Table 12.2. However, high-fidelity HMI prototypes are not suitable for complete efficiency and safety evaluation. As the screen hardware and computing platforms used in high-fidelity HMI prototypes are completely different from those in real vehicles, touch sensitivity and system response speed differ greatly from real vehicles, often making it difficult to smoothly reproduce the interaction experience in real vehicles. In addition, voice control in high-fidelity prototypes usually does not connect the same backend system as in real vehicles.

Participants for testing high-fidelity HMI prototypes can include a combination of real potential users and experts. Real users can articulate their overall subjective feelings about each module and identify significant highlights and flaws in the prototype. By contrast, experts can more accurately pinpoint design flaws and provide suggestions for improvement and optimization.

Fig. 12.4 Closed-cabin seating buck for high-fidelity HMI prototypes (the outside is a simulated driving environment)

High-fidelity HMI prototypes can use the same seating buck as rapid concept prototypes. Alternatively, they can also use closed- or semi-closed-cabin driving simulator seating bucks. These are usually modified from real vehicles, using real vehicle bodies, doors, central console, and seats, with high-fidelity HMI prototype screens installed inside the cockpit, thus reproducing a more authentic immersive experience, as shown in Fig. 12.4. Regardless of the seating buck used, the screens should be installed as precisely as possible as their positions in the future mass-produced vehicle to examine whether the distribution of page elements is easily reachable.

12.2.3 Prototype Module /Prototype Vehicle Validation Phase

After the development of the HMI system software and hardware, the prototype module can be tested, which usually comprises the same screens, head unit, microphones, and other hardware as the mass-produced vehicle, as well as a software system running on the real head unit of the real vehicle. Installing the prototype module on a trial or mass-produced vehicle result in a prototype vehicle for automotive HMI system testing. The prototype vehicle can be driven on real roads, and its HMI system will be connected to electronic and electrical equipment and cloud backend, achieving functions that are fundamentally consistent with the mass-produced vehicle.

The testing and evaluation goals of the prototype module and prototype vehicle are to perform a complete and comprehensive validation of the HMI system's software and hardware. The prototypes should achieve all design goals, ensuring all software and hardware have good compatibility and can operate stably and efficiently. Issues discovered at this stage need to be categorized according to the importance and workload required for improvement. Significant issues with a manageable workload should be immediately rectified, ensuring completion before the vehicle is launched. Issues that are less critical and require a high workload can be considered for improvement via software updates after the vehicle's launch.

For a fully functional prototype vehicle, all evaluation indexes can be applied for testing. However, for prototype module that have not yet been installed in a vehicle, it is impossible to conduct tests on evaluation indexes related to driving, as shown in Table 12.2. If the prototype module can be installed in the vehicle in a timely manner, more effort should be put into testing the prototype vehicle, so that the testing of the prototype module can focus solely on relevant indexes in utility.

Prototype module and prototype vehicle testing is typically performed by professional testers. The focus of the evaluation in this phase is no longer to identify design highlights in the HMI system or obtain general user feedback. Professional testers should meticulously seek out minor design flaws and perform quantitative evaluations according to strict and complex standardized procedures. However, if a product did not include real potential users in the evaluations during the concept definition and UX/UI design phases, the user test can be considered in this phase. The only drawback is that numerous user suggestions may not be adopted at this stage before the vehicle is launched.

Prototype module testing can use the same seating buck as the rapid concept prototypes. If only utility-related indexes are tested, a simple test bench without driving simulation functions can also be applied. Prototype vehicle tests are primarily conducted on real vehicle driving simulators, shown in Fig. 12.5, with a few evaluation indexes also requiring field operational tests. Details regarding real-vehicle driving simulators can be found in Chap. 2, Sect. 2.2.

12.2.4 Mass-Produced Vehicle Benchmarking Phase

After the vehicle is launched, a comprehensive evaluation can be performed on the HMI system of the mass-produced vehicle, which can be benchmarked against competing products. The purpose of this evaluation is to perform a comprehensive validation of the HMI system's performance, determine its position in the current market environment, and understand its advantages and disadvantages compared with competing products to provide support for the product's marketing as well as guidance for the manufacturer's subsequent R&D.

The testing and evaluation methods for mass-produced vehicles are the same as those for prototype vehicles. They both involve all evaluation indexes, as shown in Table 12.2. The testing process primarily includes real vehicle driving simulators, and

Fig. 12.5 Real vehicle driving simulation bench in the Human-Vehicle Relationship Laboratory

a few evaluation indexes also require field operational tests. The testing and evaluation are usually performed by professional testers. The main difference between mass-produced vehicle test and prototype vehicle test is how the evaluation results are used. In prototype vehicle testing, the main purpose is to discover problems and make corrections before the product is launched; conversely, mass-produced vehicle evaluations are conducted mainly to understand the position of the product in the market instead of discovering significant product defects. The overall benchmarking process for mass-produced vehicles can also involve in-depth interviews with real potential users. However, the focus of such research is usually not the HMI system and does not ask the users to provide comprehensive and systematic feedback on the HMI system; therefore, this aspect will not be elaborated here.

12.3 Avoiding "Pretend" Digitalization

The automotive HMI evaluation system is not only useful for the evaluation of phase-specific prototypes or complete systems but also provides a framework for designers and developers in terms of product innovation. Starting from the HMI system design and R&D, it can guide the entire process in the right direction.

In recent years, intelligence has become the most important development trend for automotive cabins. Digitalization is an important means to achieve intelligence as well as the design direction for several automotive cabins. In a narrow sense, digitalization involves transforming various functions and information inside and outside the cabin into a computer code and interconnecting them for collaborative operation. For example, electronic rear-view mirrors (i.e. camera monitor system) can convert real external scenes into image signals displayed on in-vehicle screens. In a broader sense, digitalization also needs to enhance the users' perception of the vehicle's digitalization level. For instance, larger screens and narrower frames can enhance the users' perception of the vehicle's overall digitalization level.

Owing to the increasing importance of intelligence and digitalization in automotive cabins, innovative HMI designs are continually emerging. Accordingly, it is important to determine which of these designs can truly enhance the user's experience and value. Furthermore, it is imperative to determine which designs are digitalized for the sake of digitalization or, in other words, are "pretending" to be digitalized. Discussions on these issues can sometimes be bewildering. The automotive HMI evaluation system can provide designers and developers with two easy-to-implement principles:

- An innovative HMI design needs to significantly improve at least one of the seven first-level evaluation indexes. Otherwise, the design will not provide real value to users and, instead, may deviate from the experience, accumulating technology unnecessarily.
- An innovative HMI design may have a negative impact on certain first-level indexes while improving others. Designers should make trade-off judgments based on factors such as brand positioning and user scenarios before deciding whether to introduce the innovative design.

For example, many companies face a similar issue when defining the automotive cabin: should touch-sensitive buttons replace traditional physical buttons for adjusting the climate control temperature? Without a fixed framework, it may be challenging for the design and R&D teams to reach a consensus efficiently because they may approach the problem from different perspectives. By contrast, under the framework of the HMI evaluation system, everyone can perform the following analysis based on the seven first-level indexes: in terms of utility, touch-sensitive buttons may decrease the task success rate if they are not sensitive; however, this can be overcome by improving the hardware standard. In terms of safety and efficiency, touch-sensitive buttons do not have a pressing stroke and require more visual observation from the user to confirm the status, causing more driver distractions and longer operation times. Regarding cognition, touch-sensitive buttons do not have tactile feedback; however, this can be simulated via other stimuli (e.g., by adding vibration). From an esthetic perspective, touch buttons may enhance the sense of technology. Finally, touch buttons will not have a significant impact in terms of intelligence and value. After such an analysis, the broad question of whether to use touch buttons can be transformed into three specific and easy-to-judge questions: first, for the product's target users, can the touch-sensitive button design enhance the sense of technology and to what extent? This can be determined via research. Second, to what extent can the potential negative impact of touch buttons on safety, efficiency, and cognition be minimized? Third, to enhance the sense of technology, can the shortcomings of touch buttons in other areas be accepted? Based on these three questions, different enterprises and products can find their answers.

12.4 Future Development Directions for Automotive Experience Evaluation

The current wave of automotive intelligence is surging ahead. The automotive HMI evaluation system introduced in this book is just a starting point for automotive experience evaluation. In the future, automotive experience evaluation will continue to develop in more directions. The 3D orthogonal evaluation system framework and the definitions of each evaluation index introduced in this book can serve as references for new evaluation directions.

Autonomous driving, including an advanced driver assistance system, is another important direction for automotive intelligence that requires not only sensitive sensors, efficient algorithms, and higher computing power, but also interactions with the driver. The interactive experience of autonomous driving mainly includes three parts: the first is the control of autonomous driving functions, including phased autonomous driving initiation, setting the target vehicle speed, guiding the vehicle to change lanes, and restarting after pauses in autonomous driving. The second is to display the surrounding road conditions and the working status of autonomous driving, which includes the positions of surrounding vehicles, pedestrians, and other traffic participants, fixed obstacles, and potential hazard levels in all directions. The third is the alert and intervention system of autonomous driving in emergency situations, including omni-directional collision warnings, alerting the driver to take over as soon as possible, and automatic braking. The interactive experience of autonomous driving needs to gain sufficient trust from the driver, involve a simplified operation process, demonstrate absolute reliability of command input, avoid maloperations, and clearly display the interaction status in real time.

The interactive experience of the front and rear passengers is gaining importance in automotive HMI systems [1]. Common passenger interaction features include listening to music, watching videos, playing games, singing karaoke, searching for navigation destinations, and browsing POIs along a planned route. Passengers do not need to drive and can fully concentrate on such features when using the HMI system. Therefore, the evaluation of safety is not necessary for passenger interaction, and efficiency is not as important. As a result, the evaluation indexes for passenger interaction experiences are closer to those for other consumer electronics rather than those for driver interaction. As the driver and multiple passengers share a small space, the evaluation of passenger interaction also needs to consider the reduction of interference between vehicle occupants. For example, the front passenger display can utilize a privacy filter to prevent the driver from watching during the driving process, and the in-vehicle microphone can make use of voice source localization technology to identify which passenger is inputting voice commands.

The targets of automotive HMI are not only the vehicle occupants but also pedestrians outside the vehicle and other vehicle drivers. The vehicle can send traffic guidance information to people outside the vehicle, such as asking pedestrians to cross the road first or displaying some interesting texts and patterns [2, 3]. The external HMI information can be directly displayed through programmable lights, projected on the ground through matrix headlights, or played via external speakers. External HMI needs to allow people outside the vehicle to quickly understand the interaction information when they see it for the first time. It should also allow the driver to understand the displayed content of the external interaction at any time and conveniently turn it on and off or switch it. External HMI is still in its nascent stages, and only a small number of mass-produced vehicles have included this feature. Consequently, the path of its future development is unclear, and it is difficult to formulate comprehensive and systematic evaluation indexes for external HMI for the time being.

References

1. Ma J, Li J, Wang W, Huang H, Zhang X, Zhao J. The impact of co-pilot displays use on driver workload and driving performance exploring the impact of co-pilot display on drivers' workload and driving performance [J]. 2024, 114: 104138.
2. Métayer N, Coeugnet S. Improving the experience in the pedestrian's interaction with an autonomous vehicle: An ergonomic comparison of external HMI [J]. Applied Ergonomics, 2021, 96, 103478.
3. Papakostopoulos V, Nathanael D, Portouli E, Amditis A. Effect of external HMI for automated vehicles (AVs) on drivers' ability to infer the AV motion intention: A field experiment [J]. Transportation Research Part F: Traffic Psychology and Behaviour, 2021, 82, 32–42.

Printed by Printforce, the Netherlands